AI ETHICS

AI ETHICS

MARK COECKELBERGH

The MIT Press | Cambridge, Massachusetts | London, England

This book was set in Chaparral Pro by Toppan Best-set Premedia Limited. Printed and bound in the United States of America.

Library of Congress Cataloging-in-Publication Data

Names: Coeckelbergh, Mark, author.
Title: AI ethics / Mark Coeckelbergh.
Description: Cambridge, MA : The MIT Press, [2020] | Series: The MIT Press essential knowledge series | Includes bibliographical references and index.
Identifiers: LCCN 2019018827 | ISBN 9780262538190 (pbk. : alk. paper)
Subjects: LCSH: Artificial intelligence—Moral and ethical aspects.
Classification: LCC Q334.7 .C64 2020 | DDC 170—dc23 LC record available at https://lccn.loc.gov/2019018827

10   9   8   7   6   5   4   3   2

for Arno

CONTENTS

SERIES FOREWORD

The MIT Press Essential Knowledge series offers accessible, concise, beautifully produced pocket-size books on topics of current interest. Written by leading thinkers, the books in this series deliver expert overviews of subjects that range from the cultural and the historical to the scientific and the technical.

In today's era of instant information gratification, we have ready access to opinions, rationalizations, and superficial descriptions. Much harder to come by is the foundational knowledge that informs a principled understanding of the world. Essential Knowledge books fill that need. Synthesizing specialized subject matter for nonspecialists and engaging critical topics through fundamentals, each of these compact volumes offers readers a point of access to complex ideas.

Bruce Tidor
Professor of Biological Engineering and Computer Science
Massachusetts Institute of Technology

ACKNOWLEDGMENTS

This book not only draws on my own work on this topic but reflects the knowledge and experience of the entire field of AI ethics. It would be impossible to list all the people I have discussed with and learned from over the past years, but the relevant and fast-growing communities I know include AI researchers such as Joanna Bryson and Luc Steels, fellow philosophers of technology such as Shannon Vallor and Luciano Floridi, academics working on responsible innovation in the Netherlands and the UK such as Bernd Stahl at De Montfort University, people I met in Vienna such as Robert Trappl, Sarah Spiekermann, and Wolfgang (Bill) Price, and my fellow members of the policy-oriented advisory bodies High-Level Expert Group on AI (European Commission) and Austrian Council on Robotics and Artificial Intelligence, for example Raja Chatila, Virginia Dignum, Jeroen van den Hoven, Sabine Köszegi, and Matthias Scheutz—to name just a few. I would also like to warmly thank Zachary Storms for helping with proofreading and formatting, and Lena Starkl and Isabel Walter for support with literature search.

MIRROR, MIRROR, ON THE WALL

The AI Hype and Fears: Mirror, Mirror, on the Wall, Who Is the Smartest of Us All?

When the results are announced, Lee Sedol's eyes swell with tears. AlphaGo, an artificial intelligence (AI) developed by Google's DeepMind, just secured a 4–1 victory in the game Go. It is March 2016. Two decades earlier, chess grandmaster Garry Kasparov lost to the machine Deep Blue, and now a computer program had won against eighteen-time world champion Lee Sedol in a complex game that was seen as one that only humans could play, using their intuition and strategic thinking. The computer won not by following rules given to it by programmers but by means of machine learning based on millions of past Go matches and by playing against itself. In such a case, programmers prepare the data sets and create the algorithms, but cannot know which moves the program will come up with. The AI learns by itself. After a number of unusual and surprising moves, Lee had to resign (Borowiec 2016).

An impressive achievement by the AI. But it also raises concerns. There is admiration for the beauty of the moves, but also sadness, even fear. There is the hope that even smarter AIs could help us to revolutionize health care or find solutions for all kinds of societal problems, but also the worry that machines will take over. Could machines outsmart us and control us? Is AI still a mere tool, or is it slowly but surely becoming our master? These fears remind us of the words of the AI computer HAL in Stanley Kubrick's science fiction film *2001: A Space Odyssey*, who in response to the human command to "Open the pod bay doors" answers: "I'm afraid I can't do that, Dave.". And if not fear, there may be a feeling of sadness or disappointment. Darwin and Freud dethroned our beliefs of exceptionalism, our feelings of superiority, and our fantasies of control; today, artificial intelligence seems to deal yet another blow to humanity's self-image. If a machine can do this, what is left for us? What are we? Are we just machines? Are we *inferior* machines, with too many bugs? What is to become of us? Will we become the slaves of machines? Or worse, a mere energy resource, as in the film *The Matrix*?

The Real and Pervasive Impact of AI

But the breakthroughs of artificial intelligence are not limited to games or the realm of science fiction. AI is already happening today and it is pervasive, often invisibly embedded in our day-to-day tools and as part of complex technological systems (Boddington 2017). Given the exponential growth of computer power, the availability of (big) data due to social media and the massive use of billons of smartphones, and fast mobile networks, AI, especially machine learning, has made significant progress. This has enabled algorithms to take over many of our activities, including planning, speech, face recognition, and decision making. AI has applications in many domains, including transport, marketing, health care, finance and insurance, security and the military, science, education, office work and personal assistance (e.g., Google Duplex[1]), entertainment, the arts (e.g., music retrieval and composition), agriculture, and of course manufacturing.

AI is created and used by IT and internet companies. For example, Google has always used AI for its search engine. Facebook uses AI for targeted advertising and photo tagging. Microsoft and Apple use AI to power their digital assistants. But the application of AI is wider than the IT sector defined in a narrow sense. For example, there are many concrete plans for, and experiments with, self-driving cars. This technology is also based on AI. Drones

AI is already happening today and it is pervasive, often invisibly embedded in our day-to-day tools.

use AI, as do autonomous weapons that can kill without human intervention. And AI has already been used in decision making in courts. In the United States, for example, the COMPAS system has been used to predict who is likely to re-offend. AI also enters domains that we generally consider to be more personal or intimate. For example, machines can now read our faces: not only to identify us, but also to read our emotions and retrieve all kinds of information.

The Need to Discuss Ethical and Societal Problems

AI can have many benefits. It can be used to improve public and commercial services. For example, image recognition is good news for medicine: it can help with the diagnosing of diseases such as cancer and Alzheimer. But such everyday applications of artificial intelligence also show how the new technologies raise ethical concerns. Let me give some examples of questions in AI ethics.

Should self-driving cars have built-in ethical constraints, and if so, what kind of constraints, and how should they be determined? For example, if a self-driving car gets into a situation where it must choose between driving into a child or into a wall to save the child's life but potentially killing its passenger, what should it choose? And should autonomous lethal weapons be allowed at all? How many decisions and how much of those decisions do we want to delegate to AI? And who is responsible when something goes wrong? In one case, the judges put more faith in the COMPAS algorithm than in agreements reached by the defense and the prosecution.[2] Will we rely too much on AI? The COMPAS algorithm is also highly controversial since research has shown that the algorithm's false positives (people who were predicted to re-offend but did not) were disproportionately black (Fry 2018). AI can thus reinforce bias and unjust discrimination. Similar problems can arise with algorithms that recommend decisions about mortgage applications and job applications. Or consider so-called predictive policing: algorithms are used to forecast where crimes are likely to occur (e.g., which area of a city) and who might commit them, but the result might be that specific socioeconomic or racial groups will be disproportionately targeted by police surveillance. Predictive policing has already been used in the United States and, as a recent AlgorithmWatch (2019) report shows, also in Europe.[3] And AI-based facial recognition technology is often used for surveillance and can violate people's privacy. It can also more or less predict sexual preferences. No information from your phone and no biometric data are needed. The machine does its work from a distance. With cameras on the street and other public spaces, we can be identified and "read," including our mood. By means of analysis of our data, our mental and bodily health can be predicted—without us knowing it. Employers can use the technology to monitor our performance. And algorithms that are active on social media can spread hate speech or false information; for example, political bots can appear as real people and post political content. A known case is the 2016 Microsoft chatbot named Tay that was designed to have playful conversations on Twitter but, when it got smarter, started to tweet racist things. Some AI algorithms can even create false video speeches, such as the video that was composed to misleadingly resemble a speech by Barack Obama.[4]

The intentions are often good. But these ethical problems are usually unintended consequences of the technology: most of these effects, such as bias or hate speech, were not intended by the developers or users of the technology. Moreover, one critical question to be asked is always: Improvement for whom? The government or the citizens? The police or those who are targeted by the police? The retailer or the customer? The judges or the accused? Questions concerning power come into play, for instance when the technology is shaped by only a few mega corporations (Nemitz 2018). Who shapes the future of AI?

This question points up the social and political significance of AI. AI ethics is about technological change and its impact on individual lives, but also about transformations in society and in the economy. The issues of bias and discrimination already indicate that AI has societal relevance. But it is also changing the economy and therefore perhaps the social structure of our societies. According to Brynjolfsson and McAfee (2014), we have entered a Second Machine Age in which machines are not only complements to humans, as in the Industrial Revolution, but also substitutes. As professions and work of all kinds will be affected by AI, our society has been predicted to change dramatically as technologies once described in science fiction enter the real world (McAfee and Brynjolfsson 2017). What is the future of work? What kind of lives will we have when AIs take over jobs? And who is the "we"? Who will gain from this transformation, and who will lose?

This Book

Based on spectacular breakthroughs, a lot of hype surrounds AI. And AI is already used in a wide range of knowledge domains and human practices. The first has given rise to wild speculations about the technological future and interesting philosophical discussions about what it means to be human. The second has created a sense of urgency on the part of ethicists and policymakers to ensure that this technology benefits us instead of creating insurmountable challenges for individuals and societies. These latter concerns are more practical and immediate.

AI ethics is about
technological change
and its impact on
individual lives, but also
about transformations
in society and in the
economy.

This book, written by an academic philosopher who also has experience with advice for policymaking, deals with both aspects: it treats ethics as related to all these questions. It aims to give the reader a good overview of the ethical problems with AI understood broadly, ranging from influential narratives about the future of AI and philosophical questions about the nature and future of the human, to ethical concerns about responsibility and bias and how to deal with real-world practical issues raised by the technology by means of policy—preferably before it is too late.

What happens when it is "too late"? Some scenarios are dystopian and utopian at the same time. Let me start with some dreams and nightmares about the technological future, influential narratives that, at least *at first sight*, seem relevant to evaluating the potential benefits and dangers of artificial intelligence.

SUPERINTELLIGENCE, MONSTERS, AND THE AI APOCALYPSE

Superintelligence and Transhumanism

The hype surrounding AI has given rise to all kinds of speculations about the future of AI and indeed the future of what it is to be human. One popular idea, which is not only repeated often in the media and in the public discourse about AI but is also entertained by influential tech people who develop AI technology such as Elon Musk and Ray Kurzweil, is that of superintelligence and, more generally, the idea that machines will take over, will master us rather than the other way around. For some, this is a dream; for many, a nightmare. And for some, it is both at the same time.

The idea of superintelligence is that machines will surpass human intelligence. It is often connected with the idea of an intelligence explosion and a technological singularity. According to Nick Bostrom (2014), our predicament will be comparable to that of gorillas, whose fate is today entirely dependent on us. He sees at least two paths to superintelligence and what is sometimes called an *intelligence explosion*. One is that AI will develop recursive self-improvement: an AI could design an improved version of itself, which in turn designs a smarter version of itself, and so on. Another path is whole brain emulation or uploading: a biological brain that could be scanned, modeled, and reproduced in and by intelligence software. This simulation of a biological brain would then be connected to a robot body. Such developments would then lead to an explosion of nonhuman intelligence. Max Tegmark (2017) imagines that a team could create an AI that will become all-powerful and run the planet. And Yuval Harari writes about a world in which humans no longer dominate but worship data and trust algorithms to make their decisions. After all humanist illusions and liberal institutions are destroyed, humans dream only of merging into the data flow. The AI follows its own path, "going where no human has gone before—and where no human can follow" (Harari 2015, 393).

The idea of an intelligence explosion is closely related to that of the *technological singularity*: a moment in human history when exponential technological progress would bring such a dramatic change that we no longer comprehend what happens and "human affairs as we understand them today came to an end" (Shanahan 2015, xv). In 1965, the British mathematician Irving John Good speculated about an ultraintelligent machine that designs better machines; in the 1990s, science fiction author and computer scientist Vernor Vinge argued that this would mean the end of the human era. Computer science pioneer John von Neumann already suggested the idea in the 1950s. Ray Kurzweil (2005) embraced the term "singularity" and predicted that AI, together with computers, genetics, nanotechnology, and robotics, will lead to a point when machine intelligence will be more powerful than all human intelligence combined, and when, ultimately, human and machine intelligence will merge. Humans will transcend the limitations of their biological bodies. And as the title of his book states: the singularity is near. He thinks it's going to happen around 2045.

This story need not have a happy ending: for Bostrom, Tegmark, and others, "existential risks" are attached to superintelligence. The result of these developments may be that a superintelligent AI takes over and threatens human intelligent life. Whether such an entity would be conscious or not, and more generally whatever its status or how it comes into being, the worry here is about what the entity would do (or not do). The AI may not care about our human goals. Having no biological body, it would not even understand human suffering. Bostrom offers the thought experiment of an AI that is given the goal of maximizing the manufacture of paperclips, which it does by converting the Earth and the humans who live on it into resources for producing paperclips. The challenge for us today, then, is to make sure that we build AI that somehow does not raise this control problem—that it does what we want and takes into consideration our rights. For example, should we somehow limit the AI's capabilities? How are we to contain AI?[1]

A related cluster of ideas is *transhumanism*. In light of superintelligence and disappointment with human frailty and "errors," transhumanists such as Bostrom argue that we need to enhance the human being: make it smarter, less vulnerable to disease, live longer, and potentially even immortal—thus leading to what Harari calls the *Homo deus*: humans have been upgraded into gods. As Francis Bacon already said in "The Refutation of Philosophies": humans are "mortal gods" (Bacon 1964, 106). Why not try to achieve immortality? But even if that could not be achieved, the human machine needs an upgrade, according to transhumanists. If we don't do that, humans risk remaining "the slow and increasingly inefficient part of" AI (Armstrong 2014, 23). The human biology needs to be reengineered, and, some transhumanists argue, why not dispense with biological parts altogether and design nonorganic intelligent beings?

Although most philosophers and scientists who entertain these ideas take care to distinguish their views from science fiction and religion, many researchers interpret their ideas in precisely these terms. For a start, it is not clear how relevant their ideas are to current technological developments and AI science, and whether there is a real chance that we will get to superintelligence in the foreseeable future—if we can get to it at all. Some straightforwardly reject its very possibility (see the next chapter), and those who are prepared to accept that it is possible in principle, such as, for example, scientist Margaret Boden, do not think it is likely to happen in practice. The idea of superintelligence assumes that we will develop so-called *general artificial intelligence*, or intelligence that matches or exceeds that of humans, and there are many hurdles to get over before we achieve this. Boden (2016) has argued that AI is less promising than many people assume. And a White House report from 2016 endorses a consensus among private-sector experts that general AI will not be achieved for at least decades. Many researchers in AI also reject the dystopian views that Bostrom and others promote, and stress the positive use of AI as helper or teammate. But the question is not only what will actually happen in the future. Another concern is that this discussion about the (far-off) future impacts of AI distracts from the real and current risks of actually deployed systems (Crawford and Calo 2016). There seems to be a real risk that in the near future the systems will *not be smart enough* and that we will insufficiently understand their ethical and societal implications and nevertheless use them widely. The overemphasis on intelligence as humanity's main feature and our only ultimate goal is also questionable (Boddington 2017).

Nevertheless, ideas such as superintelligence continue to influence the public discussion. They are also likely to have an impact on technology development. For example, Ray Kurzweil is not only a futurist. Since 2012 he has been director of engineering at Google. And Elon Musk, CEO of Tesla and SpaceX and a very well-known public figure, seems to endorse the superintelligence and existential risk scenarios (doom scenarios?) from Bostrom and Kurzweil. He has repeatedly warned of the dangers of artificial intelligence, seeing it as an existential threat and claiming that we cannot control the demon (Dowd 2017). He thinks humans will probably go extinct, unless human and machine intelligence merge or we manage to escape to Mars.

Perhaps these ideas are so influential because they touch on deep concerns and hopes regarding humans and machines that are present in our collective consciousness. Whether or not one rejects these particular ideas, there are clear links to fictional narratives in human culture and history that try to make sense of the human and our relation to machines. It's worth making these narratives explicit in order to contextualize and better understand some of the ideas. More generally, it is important to incorporate narrative research into AI ethics—for example, to understand why certain narratives are prevalent, by whom they are created, and who benefits from them (Royal Society 2018). It can also help us to construct new narratives of the future of AI.

Frankenstein's New Monster

One way to get beyond the hype is to consider some relevant narratives from the history of human culture that shape the current public discussion about AI. This is not the first time that people have asked questions about the future of humanity and the future of technology. And, however exotic some ideas about AI may appear, we can explore connections with rather familiar ideas and narratives that are present in our collective consciousness, or more precisely, the collective consciousness of the West.

First, there is a long history of thinking about humans and machines or artificial creatures, in both Western and non-Western cultures. The idea of creating living beings from inanimate matter can be found in creation stories in Sumerian, Chinese, and Jewish, Christian, and Muslim traditions. The ancient Greeks already had the idea of creating artificial humans, in particular artificial women. For example, in *The Iliad*, Hephaestus is said to be assisted by servants made from gold to look like women. In the famous myth of Pygmalion, a sculptor falls in love with the ivory statue of a woman he's made. He wishes that she would come to life, and the goddess Aphrodite grants his wish: her lips become warm and her body soft. We can easily see the link here to contemporary sex robots.

These narratives come not only from myths: in his book *Automata*, the Greek mathematician and engineer Hero of Alexandria (ca. 10–ca. 70 CE) published descriptions of machines that made people in temples believe they were seeing acts of the gods; in 1901, an artifact was found in the sea, the Antikythera mechanism, which has been identified as an ancient Greek analog computer based on a complex clockwork mechanism. But fictional stories in which machines become human-like especially fascinate us. Consider, for example, the legend of the Golem: a monster made of clay created by a rabbi in the sixteenth century, which then gets out of control. Here we encounter an early version of the control problem. The myth of Prometheus is also often interpreted in this way: he steals fire from the gods and gives it to humans, but is then punished. His eternal torment is to be bound to a rock while every day an eagle eats his liver. The ancient lesson was to warn of hubris: such powers are not meant for mortals.

However, in Mary Shelley's *Frankenstein*—which has the telling subtitle *The Modern Prometheus*—the creation of intelligent life from lifeless matter becomes a modern scientific project. The scientist Victor Frankenstein creates a human-like being from the parts of corpses, but loses control over his creation. Whereas the rabbi can still control the Golem in the end, that is not so in this case. *Frankenstein* can be seen as a Romantic novel that warns of modern technology, but it is informed by the science of its day. For example, the use of electricity—then a very new technology—plays an important role: it is used to animate the corpse. It also makes references to magnetism and anatomy. Thinkers and writers at the time debated about the nature and origin of life. What is the life force? Mary Shelley was influenced by the science of her day.[2] The story shows how nineteenth-century Romantics were often fascinated by science, as much as they hoped for poetry and literature to liberate us from the darker sides of modernity (Coeckelbergh 2017). The novel should not necessarily be seen as against science and technology: the main message seems to be that scientists need to take responsibility for their creations. The monster runs away, but it does so because its creator rejects it. This lesson is important to keep in mind for the ethics of AI. Nevertheless, the novel clearly stresses the danger of technology that goes wild, in particular the danger of artificial humans running amok. This fear resurfaces in contemporary concerns about AI getting out of control.

In Mary Shelley's *Frankenstein*—which has the telling subtitle *The Modern Prometheus*—the creation of intelligent life from lifeless matter becomes a modern scientific project.

Moreover, as in *Frankenstein* and the Golem legend, a narrative of competition emerges: the artificial creation competes with the human. This narrative continues to shape our science fiction about AI, but also our contemporary thinking about technologies such as AI and robotics. Consider the 1920 play *R.U.R.* which is about robot slaves that revolt against their masters, the already-mentioned *2001: A Space Odyssey* from 1968 in which an AI starts killing the crew in order to fulfill its mission, or the 2015 film *Ex Machina*, in which AI robot Ava turns on its creator. The *Terminator* films also fit this narrative of machines turning against us. The science-fiction writer Isaac Asimov called this fear "the Frankenstein complex": fear of robots. This is also relevant to AI today. It is something scientists and investors have to deal with. Some argue against it; others help to create and sustain the fear. I've already mentioned Musk. Another example of an influential figure who has spread fear about AI is physicist Stephen Hawking, who said in 2017 that the creation of AI could be the worst event in the history of our civilization (Kharpal 2017). The Frankenstein complex is widespread and deeply rooted in Western culture and civilization.

Transcendence and the AI Apocalypse

Ideas such as transhumanism and the technological singularity have precedents or at least parallels in the history of Western religious and philosophical thinking, especially in the Judeo-Christian tradition and Platonism. In contrast to what many people think, religion and technology have always been connected in the history of Western culture. Let me limit my discussion to transcendence and apocalypse.

In theistic religion, transcendence means that a god is "above" and independent of the material and physical world, as opposed to in the world and part of the world (immanence). In the Judeo-Christian monotheistic tradition, God is seen as transcending his creation. God can also be seen at the same time as permeating all creation and beings (immanence), and, for example, in Catholic theology, God is understood as revealing himself immanently through his son (Christ) and the Holy Spirit. Frankensteinian narratives about AI seem to stress transcendence in the sense of a split or gap between creator and creation (between *Homo deus* and AI), without giving much hope that this split or gap can be bridged.

Transcendence can also refer to going beyond limits, surpassing something. In Western religious and philosophical history, this idea often took the shape of going above and beyond the limits of the material and physical

In contrast to what many people think, religion and technology have always been connected in the history of Western culture.

world. For example, in the second-century CE Mediterranean world, Gnosticism saw all matter as evil and aimed at liberating the divine spark from the human body. Earlier, Plato saw the body as the prison of the soul. In contrast to the body, the soul is seen as immortal. In his metaphysics, he distinguished between the forms, which are eternal, and the things in the world, which are changing—the former thus transcend the latter. In transhumanism, we see some ideas that are reminiscent of this. Not only does it retain the goal of transcendence in the sense of overcoming human limitations, but also the specific ways this transcendence is supposed to happen evoke Plato and Gnosticism: to reach immortality, the biological body must be transcended by means of uploading and the development of artificial agents. More generally, when AI and related science and technology use mathematics to abstract more pure forms from the messy material world, this can be interpreted as a Platonic program realized by technological means. The AI algorithm turns out to be a Platonic machine that extracts form (a model) from the (data) world of appearances.

Transcendence can also mean surpassing the human condition. In the Christian tradition, this can take the form of trying to bridge the gap between God and humans by making humans into gods, perhaps by restoring their original God-likeness and perfection (Noble 1997). But the transhumanist quest for immortality is an ancient one. It can be found already in Mesopotamian mythology: one of the oldest written tales of humanity, the *Epic of Gilgamesh* tells the story of the king of Uruk (Gilgamesh) who seeks immortality after the death of his friend Enkidu. He does not find it: he manages to pluck a plant that is said to restore youth, but a serpent steals it, and in the end, he has to learn the lesson that he must face the reality of his own death; the quest for immortality is futile. Throughout the history of humanity, people have searched for the elixir of life. Today science looks for anti-aging therapies. In this sense, the transhumanist quest for immortality or longevity is not new or exotic; it is one of the oldest dreams of humanity and an exponent of some contemporary science. In transhumanist hands, AI becomes a transcendence machine that promises immortality.

Other ancient concepts that help us to contextualize transhumanist ideas, in particular the technological singularity, are apocalypse and eschatology. The ancient Greek term *apocalypse*, which also plays a role in the Jewish and Christian world, refers to revelation. Today it often refers to the content of a particular kind of revelation: the vision of an end time or end-of-the-world scenario. In religious contexts, we find the term *eschatology*: a part of theology concerned with the final events of history and the ultimate destiny of humanity. Most apocalyptic and eschatological ideas involve a radical and often violent disruption or destruction of the world, while heading toward a new, higher reality, being, and level of consciousness. This also reminds us of so-called doomsday cults and sects, which were and are all about predicting disaster and the end of the world. While typically transhumanists have nothing to do with such religious cults and practices, clearly the idea of a technological singularity bears some resemblance to apocalyptic, eschatological, and doomsday narratives.

Thus, while the development of AI is based on a science that is supposed to be nonfictional and secularized, and while transhumanists usually distance themselves from religion and reject any suggestion that their works are fictional, science fiction and ancient religious and philosophical ideas inevitably play a role when we discuss the future of AI in these terms.

How to Go beyond Competition Narratives and beyond the Hype

Now one may ask: is there a way out? Can we go beyond competition narratives and find more immanent ways of making sense of the future of AI and similar technologies? Or is Western thinking about AI doomed to remain in the prison of these modern fears and fascinations and their ancient roots? Can we get beyond the hype, or will the discussion remain focused on superintelligence? I think we have ways out.

While typically transhumanists have nothing to do with such religious cults and practices, clearly the idea of a technological singularity bears some resemblance to apocalyptic, eschatological, and doomsday narratives.

First, we can look beyond Western culture to find different kinds of non-Frankensteinian narratives about technology and non-Platonic ways of thinking. For example, in Japan, where technological culture is still more influenced by nature religion than in the West, in particular by the Shinto religion, and where popular culture has portrayed machines as helpers, we find a friendlier attitude toward robots and AI. Here, we find no Frankenstein complex. What is sometimes called an "animistic" way of thinking implies that AIs can also in principle have spirit or soul, can be experienced as sacred. This means that there is no narrative of competition—and no Platonic desire to transcend materiality and to constantly defend the human as being above and beyond the machine, or fundamentally different from the machine. To my knowledge, Eastern culture also has no ideas concerning an end time. In contrast to the monotheistic religions, nature religions have a cyclical understanding of time. Thus, looking beyond Western culture (or indeed to the West's own ancient past, where we also find nature religions) can help us critically evaluate the dominant narratives about the future of AI.

Second, to get beyond the hype and not limit the ethical discussion about AI to dreams and nightmares about the distant future, we can (1) use philosophy and science to critically examine and discuss the assumptions about AI and the human that play a role in these scenarios and discussions (e.g., Is general intelligence possible? What is the difference between humans and machines? What is the relation between humans and technology? What is the moral status of AI?); (2) look in more detail at what existing AI is and what it does today in various applications; (3) discuss more concrete and pressing ethical and societal problems raised by AI as it is applied today; (4) investigate AI policy for the near future; and (5) question whether the focus on AI in current public discourse is helpful in light of other problems we face, and whether intelligence should be our only focus. We will follow these paths in the next chapters.

ALL ABOUT THE HUMAN

**Is General AI Possible? Are There Fundamental
Differences between Humans and Machines?**

The transhumanist vision of the technological future assumes that general artificial intelligence (or strong AI) is
possible, but is it? That is, can we create machines with
human-like cognitive capacities? If the answer is no, then
the entire superintelligence vision is irrelevant to AI ethics. If human general intelligence is not possible in machines, we don't have to worry about superintelligence.
More generally, our evaluation of AI seems to depend on
what we think AI is and can become, and on how we think
about the differences between humans and machines. At
least since the mid-twentieth century, philosophers and
scientists have debated what computers are able to do and
become, and what the differences are between humans

and intelligent machines. Let's have a look at some of these discussions, which are as much about what *the human* is and should be as they are about what AI is and should be.

Can computers have intelligence, consciousness, and creativity? Can they make sense of things and understand meaning? There is a history of criticism and skepticism about the possibility of human-like AI. In 1972, Hubert Dreyfus, a philosopher with a background in phenomenology, published a book called *What Computers Can't Do*.[1] Since the 1960s, Dreyfus had been very critical about the philosophical basis of AI and had questioned its promises: he argued that the AI research program was doomed to fail. Before moving to Berkeley, he was working at MIT, an important place for the development of AI, which at the time was based mainly on symbolic manipulation. Dreyfus argued that the brain is not a computer and that the mind does not operate by means of symbolic manipulation. We have an unconscious background of common-sense knowledge based on experience and what Heidegger would call our "being-in-the-world," and this knowledge is tacit and cannot be formalized. Human expertise, Dreyfus argued, is based on know-how rather than know-that. AI cannot capture this background meaning and knowledge; if that's what AI aims at, it's basically alchemy and mythology. Only human beings can see what is relevant because, as embodied and existential beings, we are involved

There is a history of criticism and skepticism about the possibility of human-like AI.

in the world and are able to respond to the demands of the situation.

At the time, Dreyfus met much opposition, but later, many AI researchers would no longer promise or predict general AI. AI research moved away from reliance on symbol manipulation toward new models, including statistics-based machine learning. And while at Dreyfus's time there was still a huge gap between phenomenology and AI, today many AI researchers embrace embodied and situated cognitive science approaches, which claim to be closer to phenomenology.

That being said, Dreyfus's objections are still relevant and show how views of the human being, especially but not only in so-called continental philosophy, often clash with scientific worldviews. Continental philosophers usually stress that human beings and minds are fundamentally different from machines, and focus on (self-)conscious human experience and human existence, which cannot and should not be reduced to formal descriptions and scientific explanations. Other philosophers, however, often from the analytic tradition of philosophy, endorse a view of the human being that supports AI researchers who think that the human brain and mind *really are and work like* their computer models. Philosophers such as Paul Churchland and Daniel Dennett are good examples of the latter. Churchland thinks that science, in particular evolutionary biology, neuroscience, and AI, can fully explain human consciousness. He thinks that the brain is a recurrent neutral network. His so-called eliminative materialism denies the existence of immaterial thoughts and experiences. What we call thoughts and experiences are just brain states. Dennett too denies the existence of anything above what happens in the body: he thinks that we are "a sort of robot ourselves" (Dennett 1997). And if the human is basically a conscious machine, then such machines are possible, and not just in principle but as a matter of fact. We can try to make them. Interestingly, both continental and analytic philosophers thus argue against a Cartesian dualism that splits mind and body, but for different reasons: the first because they think that human existence is about being-in-the-world in which mind and body are not separated, the latter because for materialist reasons they think that mind is nothing separate from body.

But not all philosophers in the analytic tradition think that general or strong AI is possible. From a (later) Wittgensteinian point of view, one can argue that while a set of rules can describe a cognitive phenomenon, that doesn't imply that we actually have rules in our head (Arkoudas and Bringsjord 2014). As with Dreyfus's criticism, this at least problematizes *one kind of AI*, symbolic AI, if it assumes that this is how humans think. Another famous philosophical criticism of AI comes from John Searle, who argues against the idea that computer programs could have genuine cognitive states or understand meaning (Searle 1980). The thought experiment he offers, called the Chinese room argument, goes as follows: Searle is locked in a room and given Chinese writings but doesn't know Chinese. However, he can answer questions given to him by Chinese speakers outside the room because he uses a rulebook that enables him to produce the right answers (output) based on the documents (input) he is given. He can do that successfully without understanding Chinese. Similarly, Searle argues, computer programs can produce an output based on an input by means of rules that are given to them, but they don't understand anything. In more technical philosophical terms: computer programs don't have intentionality, and genuine understanding cannot be generated by formal computation. As Boden (2016) puts it, the idea is that meaning comes from humans.

While today's AI computer programs are often different from those Dreyfus and Searle criticized, the debate continues. Many philosophers think that there are crucial differences between how humans and computers think. For example, today one can still object that we are meaning-making, conscious, embodied, and living beings whose nature, mind, and knowledge cannot be explained away by comparisons to machines. Note again, however, that even those scientists and philosophers who believe that *in principle* there is much similarity between humans and machines, and that *in theory* general AI is possible,

We are meaning-making,
conscious, embodied,
and living beings
whose nature, mind,
and knowledge
cannot be explained
away by comparisons
to machines.

often reject Bostrom's vision of superintelligence and similar ideas that hold human-like AI to be around the corner. Both Boden and Dennett think that general AI is very difficult to realize in practice and is hence not something to worry about today.

In the background of the discussion about AI are thus deep disagreements about the nature of the human, human intelligence, mind, understanding, consciousness, creativity, meaning, human knowledge, science, and so on. If it is a "battle" at all, it is one that is as much about the human as it is about AI.

Modernity, (Post)humanism, and Postphenomenology

From a broader humanities point of view, it is interesting to contextualize these debates about AI and the human further in order to show what is at stake. They are not only about technology and the human but reflect deep divides in modernity. Let me briefly touch on three divides that indirectly shape the ethical discussions about AI. The first is an early modern divide between the Enlightenment and Romanticism. The others are relatively recent developments: one is between humanism and transhumanism, which stays within the tensions of modernity, and one is between humanism and posthumanism, which attempts to go beyond modernity.

A first way of making sense of the debate about AI and the human is to consider the tension in modernity between the *Enlightenment* and *Romanticism*. In the eighteenth and nineteenth centuries, Enlightenment thinkers and scientists challenged traditional religious views and argued that reason, skepticism, and science would show us how humans and the world really are, as opposed to how it might seem given beliefs that are unjustified by arguments and unsupported by evidence. They were optimistic about what science could do to benefit humanity. In response, Romantics argued that abstract reason and modern science had disenchanted the world and that we need to bring back the mystery and wonder that science wanted to eliminate. Looking at the debate about AI, it seems that we have not moved on much from there. Dennett's work on consciousness and Boden's work on creativity, for example, are aimed at explaining away, at "breaking the spell," as Dennett puts it. These thinkers are optimistic that science can unravel the mystery of consciousness, creativity, and so on. They react against those who resist such efforts to disenchant the human, such as continental philosophers who work in the tradition of postmodernism and stress the mystery of being human—in other words: the new Romantics. "Break the spell, or hold on to the wonders of the human being?" seems, then, a pivotal question in discussions about general AI and its future.

A second tension is between *humanists* and *transhumanists*. What is "the human," and what should the human become? Is it important to defend the human as it is, or should we revise our concept of it? Humanists celebrate the human as it is. Ethically speaking, they emphasize the intrinsic and superior value of human beings. In the debate surrounding AI, traces of humanism can be found in arguments that defend human rights and human dignity as the basis of an ethics of AI, or in the argument for the centrality of humans and their values in the development and future of AI. Here humanism often teams up with Enlightenment thinking. But it can also take more conservative or Romantic forms. Humanism can also be found in the resistance against the transhumanist project. Whereas transhumanists think we should move on to a new type of human being that is enhanced by means of science and technology, humanists defend the human as it is and stress the value and dignity of the human, which is said to be threatened by transhumanist science and philosophy.

Defensive reactions against new technologies have their own history. In the humanities and social sciences, technology has often been criticized as threatening humanity and society. Many twentieth-century philosophers, for example, were very pessimistic about science and warned against technology dominating society. But now the battle is not only about human lives and society, it is about the human itself: to enhance or not to enhance, that is the question. On the one side, the human itself becomes a scientific-technological project, open to improvement. Once the spell of the human is broken—by Darwin, neuroscience, and AI—we can get on with making it better. AI can help us to improve the human. On the other side, we should embrace the human as it is. And, some may say: what the human is always escapes us. It cannot completely be understood by science.

These tensions continue to divide the minds and hearts in this discussion. Can we get beyond them? Practically, one could give up the goal of creating human-like AI. But even then disagreements remain about the status of *AIs as models of humans* used by AI science. Do they really teach us something about how humans think? Or do they only teach us something about a particular kind of thinking, a thinking that can be formalized with mathematics, for example, or a thinking that aims at control and manipulation? How much can we really learn from these technologies about the human? Is humanity more than science can grasp? Even in more moderate discussions, the struggles about modernity surface.

To find a way out of this impasse, one could follow scholars in the humanities and social sciences who during the past fifty years have explored *nonmodern* ways of thinking. Authors such as Bruno Latour and Tim Ingold have shown that we can find less dualist, more nonmodern ways of relating to the world that go beyond the Enlightenment–Romanticism opposition. We can then try to cross the modern divide between humans and nonhumans not via modern science or transhumanism, which in their way also see humans and machines not as fundamentally opposed, but via posthumanist thinking from the (post)humanities. This brings us to the third tension: between *humanism* and *posthumanism*. Against humanists, who are accused of having done violence toward nonhumans such as animals in the name of the supreme value of the human, posthumanists question the centrality of the human in modern ontologies and ethics. According to them, nonhumans matter too, and we should not be afraid of crossing borders between humans and nonhumans. This is an interesting direction to explore, since it takes us beyond the competition narrative about humans and machines.

Posthumanists such as Donna Haraway offer a vision in which living together with machines, and even merging with machines, is seen no longer as a threat or a nightmare, as in humanism, or as a transhumanist dream come true, but as a way in which ontological and political borders between humans and nonhumans can and should be crossed. AI can then be part of not a *trans*humanist but a critical *post*humanist project, which enters from the side of humanities and the arts rather than science. Borders are crossed not in the name of science and universal progress, as some Enlightenment transhumanists may want to say, but in the name of a posthumanist politics and ideology of crossing borders. And posthumanism can also offer something else relevant to AI: it can urge us to acknowledge that *nonhumans don't need to be similar to us and should not be made similar to us*. Backed up by such a posthumanism, then, it seems that AI can free itself of the burden to imitate or rebuild the human and can explore different, nonhuman kinds of being, intelligence, creativity, and so on. AI need not be made in our image. Progress here means going beyond the human and opening ourselves up to the nonhuman to learn from it. Moreover, both transhumanists and posthumanists could agree that instead of *competing* with an AI for a given task, we could also set a common goal, which then is reached by *collaborating* and mobilizing the best humans and artificial agents can offer in order to move closer to reaching that common goal.

Another way of going beyond the competition narrative, a way that sometimes comes close to posthumanism, is an approach in philosophy of technology called *postphenomenology*. Dreyfus draws on phenomenology, in particular the work of Heidegger. But postphenomenological thinking, initiated by philosopher Don Ihde, goes beyond phenomenology of technology à la Heidegger by focusing on how humans relate to specific technologies and in particular material artifacts. This approach, often collaborating with science and technology studies, reminds

Backed up by posthumanism, AI can free itself of the burden to imitate or rebuild the human and can explore different, nonhuman kinds of being, intelligence, creativity, and so on.

us of the material dimension of AI. AI is sometimes seen as having a merely abstract or formal nature, unrelated to specific material artifacts and infrastructures. But all the formalizations, abstractions, and symbolic manipulations mentioned earlier rely on material instruments and material infrastructures. For example, as we will see in the next chapter, contemporary AI relies heavily on networks and the production of large amounts of data with electronic devices. Those networks and devices are not merely "virtual" but have to be materially produced and sustained. Moreover, against the modern subject–object divide, post-phenomenologists such as Peter-Paul Verbeek talk about the mutual constitution of humans and technology, subject and object. Instead of seeing technology as a threat, they emphasize that humans are technological (that is, we have always used technology; it is part of our existence rather than something external that threatens that existence) and that technology naturally mediates our engagement with the world. For AI, this view seems to imply that the humanist battle to defend the human against technology is misdirected. Instead, according to this approach, the human has always been technological and therefore we should rather ask *how* AI mediates humans' relation to the world and try to actively shape these mediations while we still can: we can and should discuss ethics at the stage of AI development rather than complain afterward about the problems it causes.

However, one may worry that posthumanist and post-phenomenological visions are not critical enough because they are too optimistic and too remote from scientific and engineering practice, and so insufficiently sensitive to the real dangers and ethical and societal consequences of AI. Crossing never-before-crossed borders is not necessarily unproblematic, and in practice such posthumanist and postphenomenological ideas might be of little help against the domination and exploitation we may face from technologies such as AI. One may also defend a more traditional view of the human or call for a new kind of humanism, rather than posthumanism. Thus the debate continues.

JUST MACHINES?

Questioning the Moral Status of AI: Moral Agency and Moral Patiency

One of the issues that came up in the previous chapter was whether nonhumans matter, too. Today many people think that animals matter, morally speaking. But this was not always the case. Apparently, we were wrong about animals in the past. If today many people think that AIs are just machines, are they making a similar mistake? Would superintelligent AIs, for example, deserve moral status? Would they have to be given rights? Or is it a dangerous idea to even consider the question of whether machines can have moral status?

One way of discussing what AI is and can become is to ask about the moral status of AI. Here we approach philosophical questions regarding AI, not via metaphysics, epistemology, or the history of ideas, but rather via moral philosophy. The term *moral status* (also sometimes called *moral standing*) can refer to two kinds of questions. The first concerns what the AI is capable of doing morally speaking—in other words, whether it can have what philosophers call *moral agency*, and, if so, whether it can be a full moral agent. What does this mean? It seems that the actions of AIs today already have moral consequences. Most people will agree that AI has a "weak" form of moral agency in this sense, which is similar to, say, most cars today: the latter can also have moral consequences. But given that AI is becoming more intelligent and autonomous, can an AI have a stronger form of moral agency? Should it be given or will it develop some capacity for moral reasoning, judgment, and decision making? For example: can and should self-driving cars that use AI be considered moral agents? These questions are about the ethics of AI, in the sense of *what kind of moral capacities does or should an AI have*? But questions about "moral status" can also refer to how we should treat an AI. Is an AI "just a machine," or does it deserve some form of moral consideration? Should we treat it differently than, say, a toaster or a washing machine? Would we have to confer rights upon a highly intelligent artificial entity, if such an entity were someday developed, even if it were not human? This is what philosophers call the question regarding *moral patiency*. This question is not about the ethics *by* or *in* AI but about *our*

Is an AI "just a machine"? Should we treat it differently than, say, a toaster or a washing machine?

ethics *toward* AI. Here the AI is object of ethical concern, rather than a potential ethical agent itself.

Moral Agency

Let's start with the question of moral agency. If an AI were to be more intelligent than is possible today, we can suppose that it could develop moral reasoning and that it could learn how humans make decisions about ethical problems. But would this suffice for full moral agency, that is, for human-like moral agency? The question is not entirely science fiction. If we already today hand over some of our decisions to algorithms, for example in cars or courtrooms, then it seems it would be a good thing if those decisions were morally sound. But it is not clear whether machines can have the same moral capacities as humans. They are given agency in the sense that they do things in the world, and these actions have moral consequences. For example, a self-driving car may cause an accident, or an AI may recommend sending a particular person to jail. These behaviors and choices are not morally neutral: there are clearly moral consequences for the people involved. But to deal with this problem, should AIs be given moral agency? Can they have full moral agency?

There are various philosophical positions on these questions. Some say that machines can never be moral agents at all. Machines, they argue, do not have the required capacities for moral agency such as mental states, emotions, or free will. Hence it is dangerous to suppose that they can make sound moral decisions and to totally hand over these moral decisions to them. For example, Deborah Johnson (2006) has argued that computer systems have no moral agency of their own: they are produced and used by humans, and only these humans have freedom and are able to act and decide morally. Similarly, one could say that AIs are made by humans and that hence moral decision making in technological practices should be performed by humans. On the other side of the spectrum are those who think that machines can be full moral agents in the same way that humans are. Researchers such as Michael and Susan Anderson, for example, claim that in principle it is possible and desirable to give machines a human kind of morality (Anderson and Anderson 2011). We can give AIs principles, and machines might even be better than human beings at moral reasoning since they are more rational and do not get carried away by their emotions. Against this position, some have argued that moral rules often conflict (consider, for example, Asimov's robot stories, in which moral laws for robots always get robots and humans in trouble) and that the entire project of building "moral machines" by giving them rules is based on mistaken assumptions regarding the nature of morality. Morality cannot be reduced to following rules and is not entirely a matter of human emotions—but the latter may well be indispensable for moral judgment. If general AI is possible at all, then we don't want a kind of "psychopath AI" that is perfectly rational but insensitive to human concerns because it lacks emotions (Coeckelbergh 2010).

For these reasons, we could reject the very idea of full moral agency altogether, or we could take a middle position: we have to give AIs some kind of morality, but not full morality. Wendell Wallach and Colin Allen use the term "functional morality" (2009, 39). AI systems need some capacity to evaluate the ethical consequences of their actions. The rationale for this decision is clear in the case of self-driving cars: the car will likely get into situations where a moral choice has to be made but there is no time for human decision making or human intervention. Sometimes these choices take the form of dilemmas. Philosophers talk about *trolley dilemmas*, named after a thought experiment in which a trolley barrels down a railway track and you have to choose between doing nothing, which will kill five people tied to the track, or pulling a lever and sending the trolley to another track, where only one person is tied down but is someone you know. What is the morally right thing to do? Similarly, proponents of this approach argue, a self-driving car may have to make a moral choice between, for example, killing pedestrians crossing the road and driving into a wall, thereby killing the driver. What should the car choose? It seems that we will have to make these moral decisions (beforehand) and make sure developers implement them in the cars. Or perhaps we need to build AI cars that learn from humans' choices. However, one may question whether giving AIs rules is a good way to represent human morality, if morality can be "represented" and reproduced at all, and if trolley dilemmas capture something that is central to moral life and experience. Or, from an entirely different perspective, one may ask whether humans are in fact good in making moral choices. Why imitate human morality at all? Transhumanists, for example, may argue that AIs will have a superior morality because they will be more intelligent than us.

This questioning the focus on the human leads us to another position, which does not require full moral agency and tries to leave the anthropocentric ethical position. Luciano Floridi and J. W. Sanders (2004) have argued for a mindless morality not based on properties that humans have. We could make moral agency dependent on having a sufficient level of interactivity, autonomy, and adaptivity, and on being capable of morally qualifiable action. According to these criteria, a search-and-rescue dog is a moral agent, but so is an AI web bot that filters out unwanted emails. Similarly, one could apply nonanthropocentric criteria for moral agency of robots, as proposed by John Sullins (2006): if an AI is autonomous from programmers and we can explain its behavior by ascribing moral intentions to it (like the intention to do good or harm), and if it behaves in a way that shows an understanding of its responsibility to other moral agents, then that AI is a moral agent. Thus, these views do not require full moral agency if that means human moral agency, but rather define moral agency in a way that is in principle independent of human full moral agency and the human capacities required for that. However, would such artificial moral agency be sufficient if judged by human moral standards? The practical worry is that, for example, self-driving cars may not be moral enough. The principled worry is that we stray too far from human morality here. Many people think that moral agency is and should be connected to humanness and personhood. They are not willing to endorse posthumanist or transhumanist notions.

Moral Patiency

Another controversy concerns the moral patiency of AI. Imagine that we have a superintelligent AI. Is it morally acceptable to switch it off, to "kill" it? And closer to today's AI: is it ok to kick an AI robot dog?[1] If AIs are to be part of everyday life, as many researchers predict, then such cases will inevitably come up and raise the question of how we humans should behave toward these artificial entities. But again, we do not have to look to the far-off future or to science fiction. Research has shown that already today people empathize with robots and hesitate to "kill" or "torture" them (Suzuki et al. 2015; Darling, Nandy, and Breazeal 2015), even if these robots do not have AI. Humans seem to require very little of artificial agents in order to project personhood or humanness onto them and to empathize with them. If these agents now become AI, which potentially make them more human-like (or animal-like), this seems to make the question regarding moral patiency only more urgent. For example, how should we respond to people who empathize with an AI? Are they wrong?

To say that AIs are just machines and that people who empathize with them are simply mistaken in their judgment, emotions, and moral experience is perhaps the most intuitive position. At first sight, it seems that we do not owe anything to machines. They are things, not people. Many AI researchers think along these lines. For example, Joanna Bryson has argued that robots are tools and property and that we have no obligations to them (Bryson 2010). Those who hold this position might well agree that *if* AIs were to be conscious, have mental states, and so on, we would have to give them moral status. But they will say that this condition is not fulfilled today. As we have seen in the previous chapters, some will argue that it can never be fulfilled; others think that it could be fulfilled in principle, but that this will not happen any time soon. But the upshot for the question regarding moral status is that today and in the near future AIs are to be treated as things, unless proven otherwise.

One problem with this position, however, is that it neither explains nor justifies our moral intuitions and moral experiences that tell us there is *something* wrong with "mistreating" an AI, even if that AI does not have human-like or animal-like properties such as consciousness or sentience. To find such justifications, one could turn to Kant, who argued that it is wrong to shoot a dog, not because shooting a dog breaches any duties to the dog, but because such a person "damages the kindly and humane qualities in himself, which he ought to exercise in virtue of his duties to mankind" (Kant 1997). Today we tend to think differently of dogs (although not everyone and everywhere). But it seems that the argument could be applied to AIs: we could say that we owe nothing to an AI, but still should not kick or "torture" the AI because it makes us unkind to humans. One could also use a virtue ethics argument, which is also an indirect argument since it is about humans, not about the AI: "mistreating" an AI is wrong not because any harm is done to the AI, but because our moral character is damaged if we do so. It does not make us into better persons. Against this approach

Some argue that "mistreating" an AI is wrong not because any harm is done to the AI, but because our moral character is damaged if we do so.

we could argue that in the future some AIs may have intrinsic value and deserve our moral concern, provided they have properties such as sentience. An indirect duty or virtue approach does not seem to take seriously this "other" side of the moral relation. It cares only about humans. What about the AIs? But can AIs or robots be *others* at all, as David Gunkel (2018) has asked? Again, common sense seems to say: no, AIs do not have the required properties.

An entirely different approach argues that the way we question moral status is problematic. The usual moral reasoning about moral status is based on what morally relevant properties entities have—for example, consciousness or sentience. But how do we know that the AI really has particular morally relevant properties or not? Are we sure in the case of *humans*? The skeptic says we are not sure. Yet even without this epistemological certainty we still ascribe moral status to humans on the basis of appearance. This would also be likely to happen *if* AIs were to have a human-like appearance and behavior in the future. It seems that whatever is deemed to be morally *right* by philosophers, humans will anyway ascribe moral status to such machines and, for example, give them rights. Moreover, if we look more closely at how humans *actually* ascribe moral status, it turns out that, for example, existing social relations and language play a role. For example, if we treat our cat kindly, this is not because we engage in moral reasoning about our cat, but because we already have a kind of social relation with it. It is already a pet and companion before we do the philosophical work of ascribing moral status—if we ever felt the need for such an exercise at all. And if we give our dog a personal name, then—in contrast to the nameless animals we eat—we have already conferred a particular moral status on it independent of its objective properties. Using such a relational and critical, nondogmatic approach (Coeckelbergh 2012), we could argue that, similarly, the status of AIs will be ascribed by human beings and will depend on how they will be embedded in our social life, in language, and in human culture.

Furthermore, since such conditions are historically variable—think again about how we used to treat and think about animals—perhaps some moral caution is needed before we "fix" the moral status of AI in general or any particular AI. And why even talk about AI in general or in the abstract? It seems that there is something wrong with the moral procedure of ascribing status: in order to judge it, we take the entity out of its relational context, and before we have the result of our moral procedure we already treat it, rather hierarchically, patronizingly, and hegemonically, as an entity we superior human judges will make decisions about. It seems that before we do our actual reasoning about its moral status, we have already positioned it and perhaps even done violence to it by treating it as the object of our decision making, setting up ourselves as central, powerful, and all-knowing gods on Earth who reserve the right to confer moral status upon other entities. We have also made all situational and social contexts and conditions invisible. As in the trolley dilemma case, we have reduced ethics to a caricature. With such reasoning, moral philosophers seem to do what Dreyfusian philosophers accused symbolic AI researchers of doing: formalizing and abstracting a wealth of moral experience and knowledge at the cost of leaving out what makes us human and—in addition—at the risk of begging the very question of the moral status of nonhumans. Regardless of what the actual moral status of AIs "is," as if this could be defined entirely independent from human subjectivity, it is worth critically examining our own moral attitude and the project of abstract moral reasoning itself.

Toward More Practical Ethical Issues

As the discussions in this and the previous chapter show, thinking about AI not only teaches us something about AI. It also teaches us something about ourselves: about how we think and how we actually do and should relate to nonhumans. If we look into the philosophical foundations of AI ethics, we see deep disagreements about the nature and future of humanity, science, and modernity. Questioning AI opens up an abyss of critical questions about human knowledge, human society, and the nature of human morality.

These philosophical discussions are less far-fetched and less "academic" than one may think. They will keep resurfacing when, later in this book, we consider more concrete ethical, legal, and policy questions raised by AI. If we try to tackle topics such as responsibility and self-driving cars, the transparency of machine learning, biased AI, or the ethics of sex robots, we soon find ourselves confronted with them again. If AI ethics wants to be more than a checklist of issues, it should also have something to say about such questions.

That being said, it is time now to turn to more practical issues. These concern neither the philosophical problems raised by hypothetical general artificial intelligence, nor the risks connected to superintelligence in the far future, nor other spectacular monsters of science fiction. They are about the less visible and arguably less sexy, but still very important, realities of AIs that are already in effect. AI as it already functions today does not take the role of Frankenstein's monster or the spectacular AI robots that threaten civilization, and is more than a philosophical thought experiment. AI is about the less visible, backstage but pervasive, powerful, and increasingly smarter technologies that already shape our lives today. AI ethics, then, is about the ethical challenges posed by current and near-future AI and its impact on our societies and vulnerable democracies. AI ethics is about the lives of people and it is about policy. It is about the need for us, as persons and as societies, to deal with the ethical issues *now*.

THE TECHNOLOGY

Before discussing more detailed and concrete ethical problems with AI, we have one more task to do to clear the ground: beyond the hype, we need some understanding of the technology and its applications. Leaving aside transhumanist science fiction and philosophical speculation about general AI, let's take a look at what AI technology is and does today. As the definitions of AI and other terms are themselves contested, I will not delve too deeply into philosophical discussions or historical contextualization. Here my main purpose is to give the reader an idea of the technology in question and how it is used. Let me start by saying something about AI in general; the next chapter focuses on machine learning and data science and their applications.

What Is Artificial Intelligence?

AI can be defined as intelligence displayed or simulated by code (algorithms) or machines. This definition of AI raises the question of how to define intelligence. Philosophically speaking, it is a vague concept. An obvious comparison is human-like intelligence. For example, Philip Jansen et al. define AI as "the science and engineering of machines with capabilities that are considered intelligent by the standard of human intelligence" (2018, 5). On this view, AI is about creating intelligent machines that think or (re)act like humans. However, many researchers in AI think that intelligence need not be human-like and prefer a more neutral definition that is formulated in terms independent of human intelligence and the related goals of general or strong AI. They enumerate all kinds of cognitive functions and tasks such as learning, perception, planning, natural language processing, reasoning, decision making, and problem solving—the last is also often equated with intelligence per se. For example, Margaret Boden claims that AI "seeks to make computers do the sort of things that minds can do." At first, this makes it sound like humans are the only model. However, she then enumerates all kinds of psychological skills such as perception, prediction, and planning, which are part of the "richly structured space of diverse information-processing capacities"

(2016, 1). And this information processing need not be an exclusively human affair. General intelligence, according to Boden, need not be human. Some animals can also be considered intelligent. And transhumanists dream about future minds that are no longer biologically embedded. That being said, the goal of achieving human-like capabilities and possibly human-like general intelligence has been part of AI from the beginning.

The history of AI is closely connected to that of computer science and related disciplines such as mathematics and philosophy, and hence reaches back at least to early modern times (Gottfried Wilhelm Leibniz and René Descartes, for example) if not to ancient times, with its stories about craftsmen making artificial beings and ingenious mechanical artifacts that could trick people (think of animated figures in ancient Greece or human-shaped mechanical figures in ancient China). But as a discipline on its own, AI is generally seen as having started in the 1950s, after the invention of the programmable digital computer in the 1940s and the birth of the discipline of cybernetics, defined by Norbert Wiener in 1948 as the scientific study of "control and communication in the animal and the machine" (Wiener 1948). An important moment for the history of AI was the publication of Alan Turing's 1950 paper "Computing Machinery and Intelligence" in *Mind*, which introduced the famous Turing test but was more broadly about the question whether machines can think and already speculated about machines that could learn and do abstract tasks. But the Dartmouth workshop that took place in the summer of 1956 in Hanover, New Hampshire, is generally regarded as the birthplace of contemporary AI. Its organizer John McCarthy coined the term AI, and participants included names such as Marvin Minsky, Claude Shannon, Allen Newell, and Herbert Simon. Whereas cybernetics was perceived as being too busy with analog machines, Dartmouth's AI embraced digital machines. The idea was to *simulate* human intelligence (which is not to re-create: the process is not the same as in humans). Many participants thought that a machine as intelligent as a human being would be around the corner: they expected that it would take no more than a generation.

This is the goal of *strong AI*. *Strong* or *general* AI is capable of carrying out any cognitive tasks that humans can do, whereas *weak* or *narrow* AI can perform only in specific domains such as chess, the classification of images, and so on. As of today, we have not achieved general AI and, as we have seen in the previous chapters, it is doubtful whether we ever will. Although some researchers and companies are trying to develop it, especially those who believe in the computational theory of mind, general AI is not on the horizon. Hence the ethics and policy questions in the next chapter focus on weak or narrow AI, which we already have today and which is likely to get more powerful and pervasive in the near future.

AI can be defined both as a *science* and as a *technology*. Its purpose can be to achieve a better scientific explanation of intelligence and the mentioned cognitive functions. It can help us to better understand human beings and other beings that have natural intelligence. In this way it is a science and a discipline that systematically studies the phenomenon of intelligence (Jansen et al. 2018) and sometimes the mind or brain. As such, AI is linked to other sciences such as cognitive science, psychology, data science (see below), and sometimes also neuroscience, which makes its own claims about understanding natural intelligence. But AI can also aim to develop technologies for various practical purposes, "to get useful things done," as Boden puts it: it can take the form of tools, designed by humans, that create the appearance of intelligence and intelligent behavior for practical purposes. AIs can do this by analyzing the environment (in the form of data) and acting with a significant degree of autonomy. Sometimes scientific-theoretical interests and technological purposes meet, for example in computational neuroscience, which uses tools from computer science to understand the nervous system, or in particular projects such as the European "Human Brain Project,"[1] which involves neuroscience but also robotics and AI; some of its projects combine neuroscience and machine learning in so-called big data neuroscience (e.g., Vu et al. 2018).

More generally, AI relies on and is linked to many disciplines, including mathematics (e.g., statistics), engineering, linguistics, cognitive science, computer science, psychology, and even philosophy. As we have seen, both philosophers and AI researchers are interested in understanding the mind and phenomena such as intelligence, consciousness, perception, action, and creativity. AI has influenced philosophy and vice versa. Keith Frankish and William Ramsey acknowledge this link with philosophy, stress AI's cross-disciplinarity, and combine the scientific and technological aspects in their definition of AI as "a cross-disciplinary approach to understanding, modeling, and replicating intelligence and cognitive processes by invoking various computational, mathematical, logical, mechanical, and even biological principles and devices" (2014, 1). AI is thus both theoretical and pragmatic, both science and technology. This book focuses on AI as a technology, on the more pragmatic aspect: not only because within AI the focus has shifted in this direction, but especially because it is mainly in this form that AI has ethical and societal consequences—although scientific research is also not entirely ethically neutral.

As a technology, AI can take various forms and is usually part of larger technological systems: algorithms, machines, robots, and so on. Thus, while AI may be about "machines," this term refers not only to robots, let alone only to humanoid robots. AI can be embedded in many other kinds of technological systems and devices. AI systems can take the form of software running on the web (e.g., chatbots, search engines, image analysis), but AI can also be embedded in hardware devices such as robots, cars, or "internet of things" applications.[2] For the internet of things, the term "cyber-physical systems" is sometimes used: devices that work in, and interact with, the physical world. Robots are one kind of cyber-physical system, one that directly exerts influence on the world (Lin, Abney, and Bekey 2011).

If AI is embedded in a robot, it is also sometimes called *embodied* AI. In exerting direct influence on the physical world, robotics is very dependent on physical components. But every AI, including software active on the web, "does" something and also has material aspects such as the computer on which it runs, the material aspects of the network and infrastructure it relies on, and so on. This renders problematic the distinction between, on the one hand, "virtual" web-based and "software" applications and, on the other hand, physical or "hardware" applications. AI software needs hardware and physical infrastructure to run, and cyber-physical systems are only "AI" if they are connected to the relevant software. Moreover, phenomenologically speaking, hardware and software sometimes merge in our experience and use of devices: we do not experience an interactive humanoid robot powered by AI or an AI conversational device such as Alexa as either software or hardware, but as one technological device (and sometimes as a quasi-person, e.g., Hello Barbie).

AI is likely to have a significant influence on robotics, for example, through progress in natural language processing and more human-like communication. Often these robots are called "social robots" because they are meant to participate in the daily social life of human beings, for example, as companions or assistants, by interacting with humans in a natural way. AI can thus foster further developments in social robotics.

However, regardless of the appearance and behavior of the system as a whole and its influence on its environment, which is very important phenomenologically and ethically speaking, the basis of the "intelligence" of an AI is software: an *algorithm* or a combination of algorithms. An algorithm is a set and sequence of instructions, like a recipe, that tells the computer, smartphone, machine, robot, or whatever it is embedded in, what to do. It leads to a particular output based on the information available (input). It is applied to solve a problem. To understand AI ethics, we need to understand how AI algorithms work and what they do. I will say more about this here and in the next chapter.

Different Approaches and Subfields

There are different kinds of AI. One may also say there are different *approaches* or *research paradigms*. As we saw in Dreyfus's criticism, AI was historically often *symbolic AI*. This was the dominant paradigm until the late 1980s. Symbolic AI relies on symbolic representations of higher cognitive tasks such as abstract reasoning and decision making. For example, it may decide based on a *decision tree*—a model of decisions and their possible consequences, often graphically represented as a flowchart. An algorithm that does this contains conditional statements: decision rules in the form of *if* ... (conditions) ... *then* ... (outcome). The process is a deterministic one. Drawing on a database that represents human expert knowledge, such an AI can reason through a lot of information and act as an *expert system*. It can make expert decisions or recommendations based on an extensive body of knowledge, which may be difficult or impossible for humans to read through. Expert systems are used, for example, in the medical sector for diagnosis and treatment planning. For a long time they were the most successful AI software.

Today symbolic AI is still useful, but new kinds of AI have also emerged, which may or may not be combined with symbolic AI, and which in contrast to expert systems are able to learn autonomously from data. This is done by means of an entirely different approach. The research paradigm *connectionism*, which was developed in the 1980s as an alternative to what came to be called Good Old-Fashioned Artificial Intelligence (GOFAI), and the technology of *neural networks* is based on the idea that instead of representing higher cognitive functions, we need to build interconnected networks based on simple units. Proponents claim that this is similar to how the human brain works: cognition emerges from interactions between simple processing units, called "neurons" (which, however, are not like biological neurons). Many interconnected neurons are used. This approach and technology are often used by and for *machine learning* (see the next chapter), which then is called *deep learning* if the neural networks have several layers of neurons. Some systems are hybrid; for example, DeepMind's AlphaGo is a hybrid system. Deep learning has enabled progress in fields such as machine vision and natural language processing. Machine learning that uses a neutral network can be a "black box" in the sense that while the programmers know the architecture of the network, it is not clear to others what precisely happens in its intermediate layers (between input and output) and thus how it comes to a decision. This contrasts with decision trees, which are transparent and interpretable, and can hence be checked and evaluated by human beings.

Another important paradigm in AI is one that uses more embodied and situational approaches, focusing on motor tasks and interaction rather than so-called higher cognitive tasks. The robots built by AI researchers such as Rodney Brooks of MIT do not solve problems by using symbolic representations but by interacting with the surrounding environment. For example, Brooks's humanoid robot Cog, developed in the 1990s, was built to learn by interacting with the world—as infants do. Furthermore, some people believe that mind can only arise from life; thus to create AI, we need to try to create artificial life. And some engineers take a less metaphysical and more practical approach: they take biology as a model from which to develop practical technology applications. There are also evolutionary AIs that can evolve. Some programs, using so-called genetic algorithms, can even change themselves.

This diversity in approaches to and functions of AI also implies that today AI has various *subfields*: machine learning, computer vision, natural language processing, expert systems, evolutionary computation, and so on. Today the focus is often on machine learning, but this is only one area of AI, even if these other areas are often connected to machine learning. Recently much progress has been made in computer vision, natural language processing, and the analysis of big data by means of machine learning. For example, machine learning can be used for natural language processing based on analysis of speech and written sources such as texts found on the internet. This work created the conversational agents of today. Another example is face recognition based on computer vision and deep learning, which can be used, for example, for surveillance.

Applications and Impact

AI technology can be applied in various domains (it has various applications), ranging from industrial manufacturing, agriculture, and transportation to health care, finance, marketing, sex and entertainment, education, and social media. In retail and marketing, recommender or recommendation systems are used to influence purchase decisions and to offer targeted advertising. In social media, AI can power bots: user accounts that appear to be real people but are in fact software. Such bots can post political content or chat with human users. In health care, AI is used to analyze data from millions of patients. Expert systems are also still used in this area. In finance, AI is used to analyze big data sets for market analysis and automate trading. Robot companions often include some AI. Autopilots and self-driving cars use AI. Employers can use AI to monitor employees. Video games have characters powered by AI. AIs can compose music or write news articles. It can also mimic voices of people and even create fake videos of speeches.

Given its numerous applications, AI is likely to have a pervasive impact, now and in the near future. Consider predictive policing and speech recognition, which create new possibilities for security and surveillance, peer-to-peer transportation and self-driving cars that can transform entire cities, high-frequency algorithmic trading that already shapes financial markets, or diagnostic applications in the medical sector that influence expert decision making. We should also not forget science as one of the major fields impacted by AI: by means of analysis of big data sets, AI can help scientists discover connections they would otherwise overlook. This is applicable to the natural sciences such as physics, but also to the social sciences and the humanities. AI is sure to affect the emerging field of digital humanities, for example, teaching us more about humans and about human societies.

AI also has an impact on social relations and wider societal, economic, and environmental influence (Jansen et al. 2018). AI is likely to shape human interactions and impact privacy. It is said to potentially increase bias and discrimination. It is predicted that it will lead to job losses and perhaps transform the entire economy. It could increase the gap between rich and poor and between powerful and powerless, thus accelerating injustice and inequality. Military applications may change the way wars are conducted, for example, when automated lethal weapons are used. We should also bear in mind the environmental impact, which includes the increase of energy consumption and pollution. Later I will discuss some of the ethical and societal implications in more detail, focusing on the problems and the risks of AI. But AI is also likely to have positive effects; for example, it can create new communities by means of social media, reduce repetitive and dangerous tasks by having robots take them over, improve supply chains, reduce water use, and so on.

With regard to impact—positive or negative—we should not only question the nature and extent of the impact; it is also important to ask *who* is affected and in what way by the impact. A particular impact may be more positive for some than for others. There are many stakeholders, ranging from workers, patients, and consumers, to governments, investors, and enterprises, all of whom may be affected differently. And these differences in gains and vulnerability to the impacts of AI arise not only within countries but also between countries and parts of the world. Will AI mainly benefit highly advanced and developed countries? Could it also benefit less educated and low-income people, for instance? Who will have access to the technology and be able to reap its benefits? Who will

Who will have access to
the technology and be
able to reap its benefits?
Who will be able to
empower themselves
by using AI? Who
will be excluded from
these rewards?

be able to empower themselves by using AI? Who will be excluded from these rewards?

AI is not the only digital technology that raises such questions. Other *digital information and communication technologies* also have a huge impact on our lives and societies. As we will see, some ethical problems with AI are not specific to AI. For example, there are parallels with other *automation* technologies. Consider industrial robots that are programmed and are not considered AI, but nevertheless have societal consequences when they lead to unemployment. And some of AI's problems are related to the technologies AI is connected with, such as social media and the internet, which when combined with AI present us with new challenges. For instance, when social media platforms such as Facebook use AI to learn more about their users, this raises privacy concerns.

This link to other technologies also means that often AI is not visible. This is so in the first place because it has already become an ingrained part of our everyday life. AI is often used for new and spectacular applications such as AlphaGo. But we should not forget the AI that *already* powers social media platforms, search engines, and other media and technologies that have become part of our everyday experience. AI is all over the place. The line between AI proper and other forms of technology can be blurred, rendering AI invisible: if AI systems are embedded within technology we tend not to notice them. And

We should not forget the AI that *already* powers social media platforms, search engines, and other media and technologies that have become part of our everyday lives. AI is all over the place.

if we do know that AI is involved, then it is difficult to say if it is AI that creates the problem or impact, or if it is the other technology connected to the AI. In a sense, there is no "AI" in itself: AI always relies on other technologies and is embedded in broader scientific and technological practices and procedures. While AI also raises its own specific ethical problems, any "AI ethics" thus needs to be connected to more general ethics of digital information and communication technologies, computer ethics, and so on.

Another sense in which there is no such thing as AI in itself is that the technology is always also social and human: AI is not only about technology but also about what humans do with it, how they use it, how they perceive and experience it, and how they embed it in wider social-technical environments. This is important for ethics—which is also about human decisions—and also means that it needs to include a historical and social-cultural perspective. The current media hype about AI is not the first-ever hype about advanced technologies. Before AI, "robots" or "machines" were the key words. And other advanced technologies such as nuclear technology, nanotechnology, the internet, and biotechnology have also seen a lot of debate. It is worth keeping this in mind for our discussions of AI ethics, since perhaps we can learn something from these controversies. The use and development of technology takes place in a social context. As people in technology assessment know, when technology is new it tends to be highly controversial, but once the technology becomes embedded in everyday life, the hype and the controversy deflate significantly. This is also likely to happen with AI. While such a prediction is not a good reason for abandoning the task of evaluating the ethical aspects and social consequences of AI, it helps us to see AI in context and hence to better understand it.

DON'T FORGET THE DATA (SCIENCE)

Machine Learning

Since many ethical questions about AI concern technologies that are entirely or partly based on machine learning and related data science, it is worth zooming in on this technology and science.

Machine learning refers to software that can "learn." The term is controversial: some say that what it does is not true learning because it does not have real cognition; only humans can learn. In any case, modern machine learning bears "little or no similarity to what might plausibly be going on in human heads" (Boden 2016, 46). Machine learning is based on statistics; it is a statistical process. It can be used for various tasks, but the underlying task is often pattern recognition. Algorithms can identify patterns or rules in data and use those patterns or rules to explain the data and make predictions for future data.

This is done autonomously in the sense that it happens without direct instruction and rules given by the programmer. In contrast to expert systems, which rely on human domain experts who explain the rules to programmers who then code these rules, the machine learning algorithm finds rules or patterns that the programmer has not specified. Only the objective or task is given. The software can adapt its behavior to better match the requirements of the task. For example, machine learning can help distinguish spam from significant email by going through a large number of messages and learning what counts as spam. Another example: to build an algorithm that recognizes images of cats, the programmers do not give a set of rules to the computer that define what cats are, but rather have the algorithm make its own model of cat images. It will optimize for reaching the highest prediction accuracy on a set of images of cats and non-cats. It thus aims to learn what cat images are. Humans give feedback, but they do not feed it specific instructions or rules.

Scientists used to create theories to explain data and make predictions; in machine learning, the computer creates its own models that fit the data. The starting point is the data, not the theories. In this sense, data is no longer "passive" but "active": it is "the data itself that defines what to do next" (Alpaydin 2016, 11). Researchers train the algorithm using existing data sets (e.g., old emails) and then the algorithm can predict results from new data (e.g., new emails that come in) (CDT 2018). Identifying patterns in large amounts of information (big data) is sometimes also called "data mining," in analogy to extracting valuable minerals from the earth. However, the term is misleading because the goal is the extraction of patterns from the data, the analysis of data, not the extraction of data itself.

Machine learning can be *supervised*, which means that the algorithm focuses on a particular variable that is designated as the target for prediction. For example, if the goal is to divide up people into categories (e.g., high or low security risk), the variables that predict these categories are already known, and the algorithm then learns to predict category membership (high security risk/low security risk). The programmer trains the system by providing examples and non-examples, for example, images of people that pose a high security risk and examples of people that don't. The goal is then that the system learns to predict who belongs to which category, who poses a high security risk and who not, based on new data. If the system is given enough examples, it will be able to generalize from these examples and know how to categorize new data, such as a new image of a passenger passing through airport security. *Unsupervised* means that this kind of training is not done and that the categories are not known: algorithms make their own clusters. For example, the AI makes its own security categories based on variables it selects; the programmer does not provide them. The AI may find patterns that domain experts (here: security people) have not identified yet. The categories created by the AI can look quite arbitrary to humans. Maybe they do not make sense. But statistically the categories can be identified. Sometimes they do make sense, and then this method can give us new knowledge about the categories in the real world. *Reinforcement learning*, finally, requires an indication of whether the output is good or bad. It is analogous to reward and punishment. The program is not told which actions to take but "learns" through an iterative process which actions yield reward. To take the security example: the system receives feedback from (data provided by) security people so it "knows" whether it has done a good job when it makes a particular prediction. If a person who was predicted to pose a low security risk did not cause any security problems, the system gets the feedback that its output was good and "learns" from it. Note that there is always a percentage of error: the system is never 100 percent accurate. Note also that the technical terms "supervised" and "unsupervised" have little to do with how much humans are involved in the use of the technology: while the algorithm is given some autonomy, in all three cases humans are involved in various ways.

This is also true for the data aspect of AI, including so-called big data. Machine learning based on big data has gained a lot of interest because of the availability of large amounts of data and an increase in (cheaper) computer power. Some researchers speak of a "dataquake" (Alpaydin 2016, x). We all produce data by means of our digital activities, for example when we use social media or when we buy products online. These data are of interest to commercial actors but also to governments and scientists. It has never been easier for organizations to gather, store, and process data (Kelleher and Tierney 2018). This is not only because of machine learning: the wider digital environment and other digital technologies play a role here. Online applications and social media make it easy to collect data from people. It is also less expensive to store data and computers have gotten more powerful. All this has been important for the development of AI in general, but also for data science.

Data Science

Machine learning is thus connected to *data science*. Data science aims to extract meaningful and useful patterns from data sets, and today these data sets are large. Machine learning is able to automatically analyze these large data sets. Machine learning and data science are based on statistics, which is all about going from particular observations to general descriptions. Statisticians are interested

We all produce data by means of our digital activities, for example when we use social media or when we buy products online.

in finding correlations in the data through statistical analysis. Statistical modeling looks for mathematical relationships between input and output. This is what machine learning algorithms help with.

But data science involves more than just the analysis of data by means of machine learning. The data have to be collected and prepared before they can be analyzed, and afterward the results of the analysis have to be interpreted. Data science includes challenges such as how to capture and clean data (for example, from social media and the web), how to get sufficient data, how to draw data sets together, how to restructure the data sets, how to select the relevant data sets, and what sorts of data are used. Humans thus still play an important role at all stages and with regard to all these aspects, including framing the problem, data capture, preparation of the data (the data set the algorithm trains on and the data set it will be applied to), creating or selecting the learning algorithm, interpreting the results, and deciding what action to take (Kelleher and Tierney 2018).

Scientific challenges present themselves at every stage of this process, and while the software may be easy to use, human expert knowledge is needed to deal with these challenges. Usually collaboration between humans is also needed, for example, between data scientists and engineers. Mistakes are always possible, and human choice, knowledge, and interpretation is crucial. Humans are needed here to meaningfully interpret and to direct the technology toward finding different factors and relations. As Boden (2016) remarks, AI lacks our understanding of relevance. One should add that it also lacks understanding, experience, sensitivity, and wisdom. This is a good argument why in theory and in principle humans need to be involved. But there is also an empirical argument for not leaving humans out of the picture: in practice, humans *are* involved. Without programmers and data scientists, the technology simply doesn't work. Moreover, human expertise and AI are often combined, for example, when a medical doctor uses a cancer therapy recommendation from an AI but also draws on her own experience and intuition as an expert. If human intervention is left out, things can go wrong, make no sense, or simply get ridiculous.

Take, for example, the following well-known problem from statistics, which therefore also affects the use of machine learning AI: correlations are not necessarily causal relations. Tyler Vigen's book *Spurious Correlations* (2015) gives some good examples of this. In statistics, a spurious correlation is one in which variables are not causally related but may appear to be; the correlations are due to the presence of a third, invisible factor. Examples include the correlation between the divorce rate in Maine and the per capita consumption of margarine, or the correlation between the per capita consumption of mozzarella cheese and civil engineering doctorates awarded.[1] An AI may find such correlations, but humans are needed to decide which correlations deserve further study in order to find causal relations.

Moreover, already at the stage of gathering the data and designing or creating the data set, we are making choices about how to abstract from reality (Kelleher and Tierney 2018). Abstraction from reality is never neutral, and the abstraction itself is not reality; it is a representation. This means we can discuss how good and appropriate the representation is, given a particular purpose. Compare this with a map: the map itself is not the territory, and humans have made choices in designing the map for a particular purpose (e.g., a map for car navigation versus a topographical map for hiking). In machine learning, abstraction by means of statistical methods creates a model of reality; it is not reality. It also includes choices: choices concerning the algorithm itself which provides the statistical operation that takes us from the data to the pattern/rule, but also choices involved in designing the data set the learning algorithm trains on. This choice aspect, and hence human aspect, of machine learning, means that we can and should ask critical questions about the choices being made. For example, is the training data set representative of the population? Are any biases embedded in the data? As we will see in the next chapter, these choices and issues are never mere technical questions but also have a crucial ethical component.

Applications

Machine learning and data science have numerous applications, some of which I already mentioned under the more general heading of AI. These technologies can be used to recognize faces (and even recognize emotions based on analysis of the faces), make search suggestions, drive a car, make personality predictions, predict who is going to re-offend, or recommend music to listen to. In sales and marketing, they are used to recommend products and services. For example, when you buy something on Amazon, the site will collect data about you and make recommendations on the basis of a statistical model drawing on data from all customers. Walmart has trialed face recognition technology to tackle theft in its stores; in the future it might use the same technology to determine if shoppers are happy or frustrated. The technologies also have various applications in finance. Credit reference agency Experian works with machine learning AI to analyze data about transactions and court cases in order to recommend whether or not to lend to a mortgage applicant. American Express uses machine learning to predict fraudulent transactions. In transportation, AI and big data are used to create autonomous cars. For example, BMW uses a kind of image recognition technology to analyze data that come in from the car's sensors and cameras. In health care, machine learning AI can help in the diagnosis of cancer (e.g., in analyzing radiology scans to diagnose cancer) or the detection of infectious disease. For example, DeepMind's AI analyzed one million images from eye scans and patient data, training itself to diagnose indications of degenerative eye conditions. IBM's Watson has moved beyond playing Jeopardy and is used to give recommendations for treating cancer. Wearable mobile sport and health devices also deliver data for machine learning applications. In the field of journalism, machine learning can write news stories. For example, in the UK the news agency Press Association has bots write local news pieces. AI also enters the home and private sphere, for example, in the form of robots that gather data and assistive interactive devices connected to natural language processing. Hello Barbie talks to children on the basis of natural language processing that analyzes recorded dialogues. Everything the children say is recorded, stored, and analyzed at the servers of ToyTalk. Then a response is sent to the device: Hello Barbie answers on the basis of what it has "learned" about its user. Facebook uses deep learning technologies and neural networks to structure and analyze data from the nearly two billion users of the platform who produce unstructured data. This helps the company offer targeted advertisements. Instagram analyzes the images of 800 million users in order to sell advertising to companies. Using recommendation engines that analyze customer data, Netflix is transforming itself from a distributor into a content creator: if you can predict what people want to watch, you can produce it yourself and make money with it. Data science has even been used in cooking. For example, based on analysis of nearly 10,000 recipes, IBM's Chef Watson creates its own recipes that suggest new ingredient combinations.[2] AI machine learning can also be used in education, recruiting, criminal justice, security (e.g., predictive policing), music retrieval, office work, agriculture, military weapons, and so on.

Statistics used to be seen as a not very sexy field. Today, as part of data science and in the form of AI working with big data, it is hot. It is the new magic. It is the stuff the media like to talk about. And it is big business. Some speak of a new kind of gold rush; expectations are high. Furthermore, this kind of AI is not science fiction or speculation; as the examples show, so-called narrow or weak AI is already here and it is pervasive. When it comes to its potential impact, there is nothing narrow or weak about it. It is therefore urgent to analyze and discuss the many ethical issues that are raised by machine learning and other AI technologies and their applications. This is the topic of the next chapters.

Statistics used to be seen as a not very sexy field. Today, as part of data science and in the form of AI working with big data, it is hot. It is the new magic.

PRIVACY AND THE OTHER USUAL SUSPECTS

Many ethical problems with AI are known from the area of the ethics of robotics and automation or, more generally, from the area of the ethics of digital information and communication technologies. But this by itself does not render them any less important. Moreover, because of the technology and the way it is connected to other technologies, these issues take on a new dimension and become even more urgent.

Privacy and Data Protection

Consider, for example, privacy and data protection. AI, and in particular machine learning applications working with big data, often involves the collection and use of personal information. AI can also be used for surveillance, on the street but also in the workplace and—through smart-phones and social media—everywhere. Often people do not even know that data are being gathered, or that the data they provided in one context are then used by third parties in another context. Big data also often means that data (sets) acquired by different organizations are being combined.

An ethical use of AI requires that data are collected, processed, and shared in a way that respects the privacy of individuals and their right to know what happens to their data, to access their data, to object to the collection or processing of their data, and to know that their data are being collected and processed and (if applicable) that they are then subject to a decision made by an AI. Many of these issues also arise with other information and communication technologies and, as we will see, transparency is also an important requirement in those instances as well (see later in this chapter). And data protection issues also arise in research ethics, for example, the ethics of collecting data for social science research.

If one considers the contexts in which AI is used today, however, these privacy and data protection issues become increasingly problematic. It is relatively easy to respect these values and rights when doing a survey as a social scientist: one can inform one's respondents and explicitly ask their consent, and it is relatively clear what will happen to the data. But the environment in which AI and data science are used today is usually very different. Consider social media: in spite of privacy information and applications that ask users for consent, it is unclear for users what happens to their data or even which data are collected; and if they want to use the application and enjoy its benefits, they have to consent. Often, users also don't even know that AI is powering the application they use. And often data given in one context are then moved to another domain and used for a different purpose (data repurposing), for example, when companies sell their data to other companies or move the data between different parts of the same company without users knowing this.

Manipulation, Exploitation, and Vulnerable Users

This last phenomenon also points to the risk of users being manipulated and exploited. AI is used to manipulate what we buy, which news we follow, whose opinions we trust, and so on. Researchers in critical theory have pointed to the capitalist context in which social media use happens. For example, it could be said that users of social media do free "digital labor" (Fuchs 2014) by producing data for companies. This form of exploitation can also involve AI. As social media users, we risk becoming the unpaid, exploited workforce that produces data for the AI that then analyzes our data—and in the end for the companies that use the data, which usually also include third parties. It also reminds us of Herbert Marcuse's warning in the 1960s that even so called "free," "non-totalitarian" societies have their own forms of domination, in particular the exploitation of consumers (Marcuse 1991). The danger here is that even in today's democracies, AI may lead to new forms of manipulation, surveillance, and totalitarianism, not necessarily in the form of authoritarian politics but in a more hidden and highly effective way: by changing the economy in a way that turns us all into smartphone cattle milked for our data. But AI can also be used to manipulate politics more directly, for example, by analyzing social media data to help political campaigns (as in the famous case of Cambridge Analytica, a company that used data from Facebook users without their consent for political purposes in the 2016 US presidential election), or by having bots posting political messages on social media based on analysis of people's data in terms of their political preferences in order to influence voting. Some also worry that AI, by taking over cognitive tasks from humans, infantilizes its users by "rendering them less capable of thinking for themselves or deciding for themselves what to do" (Shanahan 2015, 170). Furthermore, the risk of exploitation lies not just on the user side: AI relies on hardware that is made somewhere by people, and this production may involve the exploitation of those people. Exploitation may also be involved in the training of algorithms and the production

AI may lead to new
forms of manipulation,
surveillance, and
totalitarianism, not
necessarily in the form
of authoritarian politics
but in a more hidden
and highly effective way.

of data that are used for and by AI. AI may make life easier for its users, but not necessarily for those who mine the minerals, deal with the e-waste, and train the AI. For example, Amazon Echo's Alexa not only creates a user who does free labor, becomes a resource for data, and is sold as a product; a world of human labor is also hidden behind the scenes: miners, workers on ships, click workers who label data sets, all in the service of capital accumulation by very few people (Schwab 2018).

Some users of AI are also more vulnerable than others. Theories of privacy and exploitation often assume that the user is an autonomous and relatively young and healthy adult human being with full mental capacities. The real world is one populated with children, elderly people, people who do not have "normal" or "full" mental capabilities, and so on. These vulnerable users are more at risk. Their privacy can often be easily violated or they can be easily manipulated; and AI provides new opportunities for such violations and manipulations. Consider young children who chat with a doll that is connected to a technological system that includes AI: most likely, the child does not know that AI is being used or that data are collected, let alone what is being done with her or his personal information. An AI-powered chatbot or doll not only can collect a lot of personal information about the child and its parents in this way; it can also manipulate the child by using language and voice interface. As AI becomes part of the "internet of toys" (Druga and Williams 2017) and the internet of (other) things, this is an ethical and a political problem. The ghost of totalitarianism returns once more: not in dystopian science fiction stories or seemingly outdated postwar nightmares, but in consumer technology that is already on the market.

Fake News, the Danger of Totalitarianism, and the Impact on Personal Relationships

AI may also be used to produce hate speech and false information, or to create bots that appear to be people but in fact are AI software. I already mentioned the chatbot Tay and the fake speech of Obama. This may lead to a world in which it is no longer clear what is true and what is false, where facts and fiction mix. Whether or not this should be called "post-truth" (McIntyre 2018), these applications of AI clearly contribute to the problem. Of course, false information and manipulation existed before AI. Film, for example, has always created illusions, and newspapers have spread propaganda. But with AI, combined with the possibilities and environment of the internet and digital social media, the problem seems to increase in intensity. There seem to be more opportunities for manipulation, putting critical thinking at risk. All this reminds us once more about the dangers of totalitarianism, which benefits from confusion about the truth and in which fake news is created for ideological purposes.

However, even in a libertarian utopia things may not look so bright. False information erodes trust and thereby damages the social fabric. Overuse of technology can lead to less contact, or at least less meaningful contact, between people. Sherry Turkle (2011) has made this claim with regard to technologies such as computers and robots: we end up expecting more from technology but less from each other. This argument could also be made with regard to AI: the worry is that AI, in the form of social media or in the form of digital "companions," gives us the illusion of companionship but unsettles true relationships with friends, lovers, and families. Although this concern was already there before AI and tends to surge with every new medium (reading the newspaper or watching TV instead of talking), the argument could be that now, with AI, the technology is much *better* in creating the illusion of companionship and that this increases the risk of loneliness or deteriorating personal relationships.

Safety and Security

There are also more visible dangers. AIs, especially when embedded in hardware systems that operate in the physical world, also need to be *safe*. Consider, for example, industrial robots: they are supposed not to harm workers. Yet sometimes accidents happen in factories. Robots can kill, even if this is relatively rare. However, with AI robots, the safety problem becomes more challenging: such robots may be able to work more closely with humans, and may be able to "intelligently" avoid harming humans. But what exactly does this mean? Should they move more slowly when near a human, which slows down the process, or is it OK to move at high speed in order to do the work efficiently? There is always the risk that something could go wrong. Should the ethics of safety be a matter of trade-offs? AI robots in a home environment or in public spaces also cause safety issues. For example, should a robot always avoid bumping into humans or is it OK if it sometimes obstructs a person in order to reach its goal? These are not mere technical issues but have an ethical component: it is an issue of human lives and values such as freedom and efficiency. They also raise responsibility problems (more on this below).

Another problem that was already there before AI entered the stage, but which deserves renewed attention, is *security*. In a networked world, every electronic device or software can be hacked, invaded, and manipulated by people with malicious intentions. We all know about computer viruses, for example, which can mess up your computer. But when equipped with AI, our devices and software can do more, and when they gain more agency and this has real-world physical consequences, the security problem becomes much larger. For example, if your AI-powered self-driving car is hacked, you have more than just a "computer problem" or "software problem"; you may die. And if the software of a critical infrastructure (internet, water, energy, etc.) or a military device with lethal capacity is hacked, an entire society will likely be disrupted and many people will be harmed. In military applications, the use of autonomous lethal weapons poses an obvious security risk, especially of course to those who are targeted by them (usually not people in the West) but also to those who deploy them: they can always be hacked and turned against you. Moreover, an arms race involving these weapons could lead to a new world war. And one does not need to look far into the future: if today (non-AI) drones can already handicap a big London airport, it is not difficult to imagine how vulnerable our daily infrastructures are and how easily the maleficent use or hacking of AI could cause massive disruptions and destructions. Note also that, in contrast to, say, nuclear technology, using existing AI technology does not require expensive equipment or a long training; the hurdle for using AI for malicious purposes is thus rather low.

More mundane security problems with cars and infrastructures such as airports also remind us that while some people are more vulnerable than others, we are *all* vulnerable in the light of technologies such as AI because, as their

In a networked world,
every electronic
device or software
can be hacked, invaded,
and manipulated by
people with malicious
intentions.

agency increases and we delegate more tasks to them, we all become more dependent on them. Things can always go wrong. The new technological vulnerabilities, then, are never merely technological; they also become our human, existential vulnerabilities (Coeckelbergh 2013). The ethical problems discussed here can thus be seen as human vulnerabilities: technological vulnerabilities ultimately transform our existence as humans. To the extent that we become dependent on AI, AI is more than a tool we use; it becomes part of how we are, and how we are at risk, in the world.

Increased agency of AI, especially when it *replaces* human agency, also raises another ever more urgent ethical problem: responsibility. This is the topic of the next chapter.

A-RESPONSIBLE MACHINES AND UNEXPLAINABLE DECISIONS

How Can and Should We Attribute Moral Responsibility?

When AI is used to make decisions for us and to do things for us, we encounter a problem that is shared with all automation technologies but which becomes even more important when AI enables us to delegate far more to machines than we used to: responsibility attribution.[1] If AI is given more agency and takes over what humans used to do, how do we then attribute moral responsibility? Who is responsible for the harms and benefits of the technology when humans delegate agency and decisions to AI? To put it in terms of risk: who is responsible when something goes wrong?

When humans are doing things and making decisions, we normally connect agency with *moral responsibility*. You are responsible for what you do and for what you decide.

If AI is given more
agency and takes over
what humans used
to do, how do we
then attribute moral
responsibility?

If you have effect on the world and on others, you are responsible for those consequences. According to Aristotle this is the first condition for moral responsibility, the so-called control condition: in the *Nicomachean Ethics* he argues that the action must have its origin in the agent. This view also has a normative side: if you have agency and if you can decide, you *should* take responsibility for your actions. What we want to avoid, morally speaking, is someone who has agency and power but no responsibility. Aristotle also added another condition for moral responsibility: you are responsible if you know what you're doing. This is an *epistemic* condition: you need to be aware of what you are doing and know what the consequences could be. What we need to avoid here is someone who does things without knowing what she is doing, potentially resulting in harmful consequences.

Now let's see how these conditions fare when we delegate decisions and actions to AI. The first problem is that an AI can take actions and make decisions that have ethical consequences, but is not aware of what it does and not capable of moral thought and hence cannot be held morally responsible for what it does. Machines can be agents but not *moral* agents since they lack consciousness, free will, emotions, the capability to form intentions, and the like. For example, on an Aristotelian view, only humans can perform voluntary actions and deliberate about their actions. If this is true, the only solution is to make humans responsible for what the machine does. Humans then delegate agency to the machine, but retain the responsibility. In our legal systems, we already do this: we do not hold dogs or small children responsible for their actions, but put the legal responsibility on the shoulders of their caretakers. And in an organization, we may delegate a particular task to an individual but ascribe the responsibility to the manager in charge of the overall project—although in this case there is still some responsibility on the part of the delegate.[2] So why not let the machine perform the actions and keep the responsibility on the side of the human? This seems the best way forward, since algorithms and machines are a-responsible.

However, this solution faces several problems in the case of AI. First, an AI system may make its decisions and actions very quickly, for example, in high-frequency trading or in a self-driving car, which gives the human too little time to make the final decision or to intervene. How can humans take responsibility for such actions and decisions? Second, AIs have histories. When the AI does things in a particular context of application, it may no longer be clear who created it, who used it first, and how the responsibility should be distributed among these different parties involved. For example, an AI algorithm made in the context of a scientific project at university may find its first application in the lab at university, then in the health care sector, and later in a military context. Who is responsible? It may be difficult to track all the humans involved in the history of the particular AI and indeed in the causal history that led to a particular ethically problematic outcome. We do not always know all the people involved at the moment when a responsibility problem arises. An AI algorithm often has a long history, involving many people. This leads us to a typical problem with responsibility attribution for technological actions: there are usually many hands and, I will add, many things.

There are *many hands* in the sense that many people are involved in technological action. In the case of AI, it begins with the programmer, but we also have the end user and others. Consider the example of the self-driving car: there is the programmer, the user of the car, the owners of the car company, the other users of the road, and so on. In March 2018 an Uber self-driving car caused an accident in Arizona that resulted in the death of a pedestrian. Who is responsible for this tragic outcome? It could be those who programmed the car, those at the car company responsible for the product development, Uber, the car user, the pedestrian, the regulator (e.g., the State of Arizona), and so on. It is not clear who is responsible. It may be that responsibility cannot and should not be attributed to one person; more than one person may be responsible. But then it is not clear how to distribute the responsibility. Some people may be more responsible than others.

There are also *many things*, in the sense that the technological system consists of many interconnected elements; there are usually many components of the system involved. There is the AI algorithm, but this algorithm interacts with sensors, uses all kinds of data, and interacts with all kinds of hardware and software. All these things have histories and are connected to the people who programmed or produced them. When something goes wrong, it is not always clear if it is "the AI" that caused the problem or some other component of the system—or even where the AI ends and the rest of the technology begins. This also makes it difficult to ascribe and distribute responsibility. Consider also machine learning and data science: as we have seen, there is not only the algorithm, but also a process that includes various stages such as the collection and treatment of data, the training of the algorithm, and so on—all involving various technical elements and requiring human decisions. Again, there is a causal history that involves many humans and parts; this renders responsibility attribution difficult.

To try to deal with these issues, one could learn from legal systems or look at how insurance works; I will say something about legal notions in the policy chapters. But behind these legal and insurance systems loom more general questions about the agency of AI and responsibility for AI: how dependent do we want to be on automation technology, can we take responsibility for something the AI does, and how can we attribute and distribute responsibilities? For example, the legal notion of negligence is about whether one exercised a duty of care. But what does this duty mean in the case of AI, given that it is so hard to predict all potential ethically relevant consequences?

This brings us to the next issue. Even if the control problem could be solved, there is also the second condition for responsibility, which concerns a problem of knowledge. To be responsible, you need to know what you are doing and bringing about, and, in retrospect, know what you have done. Moreover, this issue has a relational aspect: in the case of humans, we expect that someone can explain what she has done or decided. Responsibility then means answerability and explainability. If something goes wrong, we want an answer and an explanation. For example, we ask a judge to explain her decision, or we ask a criminal why she did what she did. These conditions become very problematic in the case of an AI. First, in principle the AI of today does not "know" what it is doing, in the sense that it is not conscious and hence not aware of what it is doing and what it is bringing about. It can log and record what it does, but it does not "know what it is doing" in the same way as humans, who as conscious beings are aware of what they do and can (following Aristotle again) deliberate and reflect on what their actions and the consequences of those actions. When these conditions are not met in the case of humans, in the case of very young children, for example, we don't hold them responsible. Usually we also do not hold animals responsible.[3] If AI does not meet these conditions, then, we cannot hold an AI responsible. The solution is again to hold humans responsible for what the AI does, assuming that *they* know what the AI is doing and what they are doing with the AI, and—keeping in mind the relational aspect—that *they* are answerable for its actions and can explain what the AI did.

However, whether this assumption holds is not as straightforward as it might seem at first sight. Usually programmers and users know what they want to do with the AI, or more precisely: they know what they want the AI to do for them. They know the goal, the end; this is why they delegate the task to the AI. They might also know how the technology works in general. But, as we will see, they do not *always* know exactly what the AI is doing (at any moment in time) and cannot *always* explain what it did or how it came to its decision.

Transparency and Explainability

Here we encounter the problem of *transparency* and *explainability*. With some AI systems, the way the AI comes to its decision is clear. If the AI uses a decision tree, for example, the way it reaches its decision is transparent. It has been programmed in a way that determines the decision, given a particular input. Humans can thus explain how the AI came to its decision and the AI can be "asked" to "explain" its decision. Humans can then take responsibility for the decision or, perhaps more accurately, can make a decision based on the decision recommendation made by the AI. However, with some other AI systems, notably AIs that use machine learning and in particular deep learning that uses neural networks, this explanation and this kind of decision making is no longer possible. It is no longer transparent how the AI comes to its decision, and humans cannot fully explain the decision. They know how their system works, in general, but cannot explain a particular decision. Think about chess with deep learning: the programmers know how the AI works, but the precise way the machine arrives at a particular move (i.e., what happens in the layers of the neutral net) is not transparent and cannot be explained. This *is* a problem for responsibility, since the humans who create or use the AI cannot explain a particular decision and hence fail to know what the AI is doing and cannot answer for its actions. In one sense, the humans know what the AI is doing (for example, they know the code of the AI and know how it works in general), but in another sense they don't know (they cannot explain a particular decision), with the result that people affected by the AI cannot be given precise information about what made the machine arrive at its prediction. Thus, while all automation technology raises problems of responsibility, here we encounter a problem specific to some kinds of AI: the so-called *black box* problem.

Furthermore, even the assumption that in such cases humans have knowledge about the AI in general and of the code is not always true. Probably the initial programmers know the code and how everything works (or at least know the part they programmed), but that does not mean that the subsequent programmers and users who change or use the algorithm for specific applications fully know what the AI is doing. For example, someone using a trading algorithm may not fully understand AI, or users of social media may not even know that AI is being used, let alone understand it. And from their side, the (initial) programmers may not know the precise *future* use of the algorithm they develop or the *different domains of application* in which the algorithm may be used, let alone all the *unintended* effects of the future use of their algorithm. So even regardless of the particular problem with (deep) machine learning, there is a knowledge problem with AI to the extent that many people who use it don't know what they are doing, since they don't know what the AI is doing, what its effects are, or even *that* it is used. This, too, is a problem for responsibility and hence a serious ethical problem.

Sometimes these problems are put in terms of trust: a lack of transparency leads to less trust in the technology and in the people who use the technology. Some researchers therefore ask how we can increase trust in AI, and identify transparency and explainability as one of the factors that can increase trust, as well as, for example, avoiding bias (Winikoff 2018) or "Terminator" images of AI (Siau and Wang 2018). And as we will see in the next chapter, AI policy also often aims at building trust. However, terms such as "trustworthy" AI are controversial: should we reserve the term "trust" for talking about human–human relations, or is it fine to use it for machines as well? AI researcher Joanna Bryson (2018) has argued that AI is not a thing to be trusted but a set of software development techniques; she thinks that the term "trust" should be reserved for people and their social institutions. Moreover, the issue of transparency and explainability makes us wonder again about what kind of society we want. Here the danger is not only manipulation and domination by capitalists or technocratic elites, creating a highly divided society. The further and perhaps deeper danger that looms here is a high-tech society in which even those elites no longer know what they are doing, and in which nobody is answerable for what is happening.

As we will see, policymakers sometimes propose "explainable AI" and a "right to explanation." Yet it is questionable if it is *possible* to always have transparent AI. It seems easy to reach with classic systems. But if with contemporary machine learning applications it seems impossible in principle to explain every step of the decision process and explain decisions relating to specific individuals, we have a problem. Is it possible to "open the black box"? This would probably be a good thing, not only for ethics but also for improving the system (i.e., the model) and learning from it. For example, if the system is more explainable, and if the AI uses what we consider to be inadequate features, then humans can spot these issues and help to eliminate the spurious correlations. And if an AI identifies new strategies of playing a game and makes this more transparent to humans, then humans can learn from the machine in order to get better at playing the game. This is useful not only for gaming, but also in domains such as health care, criminal justice, and science. Some researchers therefore try to develop techniques for opening the black box (Samek, Wiegand, and Müller 2017). But if this is not yet possible or only possible to a limited extent, how do we proceed? Is the ethical issue then about trade-offs between performance and explainability (Seseri 2018)? If the cost of a well-performing system is a lack of transparency, should we still use it, or not? Or should we try to avoid this problem and find different technical solutions, so that even very advanced AIs are able to explain themselves to humans? Can we train machines to do that?

Moreover, even if transparency were desirable *and* possible, it may be difficult to realize it in practice. For example, private companies may not be willing to reveal their algorithms because they want to protect their commercial interests. Intellectual property legislation protecting those interests may also hinder this. And, as we will see in later chapters, if AI is in the hands of powerful corporations, this raises the question of who makes and who should make the AI rules.

Note, however, that ethically speaking transparency and explainability are not necessarily and certainly not only about disclosing the software code. The issue is mainly about explaining *decisions* to people. It is not primarily about explaining "how it works" but about how I, as a human being who is expected to be accountable and act responsibly, can explain my decision. How the AI works and came to its recommendation can be part of that explanation. Furthermore, disclosing a code does not by itself necessarily give knowledge about how the AI works. This depends on people's educational background and skills. If they lack the relevant technical expertise, a different kind of explanation is needed. This not only reminds us of the problem of education but also leads to the question of *what kind of explanation* is needed and, ultimately, what an explanation is.

Thus the issue about transparency and explainability also raises interesting philosophical and scientific questions, such as questions about the nature of explanation (Weld and Bansal 2018). What constitutes a good explanation? What is the difference between explanations and reasons, and can machines provide any of these? And how do humans in fact make decisions? How do they explain their decisions? There is research on this in cognitive psychology and cognitive science, which could be used for thinking about explainable AI. For example, people generally do not provide complete causal chains; instead, they select explanations, and they respond to what they believe are the explainee's beliefs: explanations are social (Miller 2018). And perhaps we also expect different explanations from machines than from humans, who, for example, often make excuses for their actions because of emotions. But if we do so, does this mean that we hold machine decision making to a higher standard than human decision making (Dignum et al. 2018), and if so, should we do so? Some researchers speak of reasoning rather than explanation. Winikoff (2018) even demands "value-based reasoning" from AIs and other autonomous systems, which should be able to represent human values and reason using human values. But can a machine "reason," and in what sense can a technological system "use" or "represent" values at all? What kind of knowledge does it have? Does it have knowledge at all? Does it have understanding at all? And, as Boddington (2017) asks, can humans necessarily fully articulate their most fundamental values?

Such problems are interesting for philosophers, but they also have direct ethical relevance and are very real and practical. As Castelvecchi (2016) puts it: opening the black box is a problem in the real world. For example, banks should explain why they deny a loan; judges should explain why they send someone (back) to prison. Explaining decisions is not only a part of what humans naturally do when they communicate (Goebel et al. 2018), it is also a moral requirement. Explainability is a necessary condition for responsible and accountable behavior and decisions. It seems required for any society that wants to take human beings seriously as autonomous and social individuals who try to act and decide responsibly and who rightly demand reasons and explanations for decisions that affect them. Whether or not AI can *directly* provide those reasons and explanations, *humans* should be able to answer the question: "Why?" The challenge for AI researchers is to ensure that if an AI is used for decision making at all, the technology is built in such a way that humans will be able as much as possible to answer that question.

BIAS AND THE MEANING OF LIFE

Bias

Another problem that is both ethical and societal, and also specific to data science–based AI as opposed to other automation technologies, is the issue of bias. When an AI makes—or more precisely *recommends*—decisions, bias may arise: the decisions may be unjust or unfair to particular individuals or groups. Although bias may also arise with classic AI—say, an expert system using a decision tree or database that contains bias—the issue of bias is often connected to machine learning applications. And while problems of bias and discrimination have always been present in society, the worry is that AI may perpetuate these problems and enlarge their impact.

Bias is often unintentional: the developers, users, and other people involved such as the management of the

While problems of bias and discrimination have always been present in society, the worry is that AI may perpetuate these problems and enlarge their impact.

company often do not foresee the discriminatory effects against certain groups or individuals. This can be because they don't understand the AI system well enough, are not sufficiently aware of the problem of bias or indeed of their own biases, or more generally do not sufficiently imagine and reflect on the potential unintended consequences of the technology and are out of touch with some relevant stakeholders. This is problematic since biased decisions can have severe consequences, for example, in terms of access to resources and freedoms (CDT 2018): individuals might not get a job, might not get credit, might end up in prison, or might even experience violence against them. And not only individuals may suffer; entire communities might be affected by the biased decisions, for example, when an entire area of the city or all people with a particular ethnic background are profiled by the AI as posing a high security risk.

Consider again the example of the COMPAS algorithm mentioned in the first chapter, which predicts if defendants are likely to re-offend and was used by judges in Florida in sentencing decisions, for example, about when someone will be granted parole. According to a study by online newsroom ProPublica, the algorithm's false positives (defendants predicted to re-offend but who actually did not) were disproportionately black, and the false negatives (defendants predicted not to re-offend but who actually re-offended) were disproportionately white (Fry

2018). Critics thus argued that there was a bias against black defendants. Another example is PredPol, a so-called predictive policing tool that has been used in the United States to predict the probability of crime in particular areas of cities and to recommend the allocation of police resources (e.g., where police officers should patrol) on the basis of these predictions. Here the worry was that the system would be biased against poor and colored neighborhoods or that disproportionate police surveillance would break down trust between people in those areas, turning the prediction of crime into a self-fulfilling prophecy (Kelleher and Tierney 2018). But bias is not only about criminal justice or policing; it can also mean, for example, that users of internet services are discriminated against if the AI profiles them unfavorably.

Bias may arise in a number of ways at all stages of design, testing, and application. To focus on design: bias can arise in the selection of the training data set; in the training data set itself, which may be unrepresentative or incomplete; in the algorithm; in the data set the algorithm is given once it is trained; in decisions based on spurious correlations (see the previous chapter); in the group that creates the algorithm; and in wider society. For example, a data set may not be representative of the population (e.g., it may be based on American white males) but still used to predict for the entire population (males and females of various ethnic backgrounds). Bias can also concern differences between countries. Many deep neural networks for image recognition are trained on the annotated data set ImageNet, which contains a disproportionate amount of data from the United States, whereas countries such as China and India, which represent a much larger part of the world's population, contribute only a small fraction (Zou and Schiebinger 2018). This can lead to cultural bias embedded in the data set. More generally, data sets may be incomplete or of poor quality, which can lead to bias. The prediction might also be based on too little data, for example, in the case of murder prediction: there are not *that* many murders, which renders generalization problematic. Another example: some researchers worry about the lack diversity in the AI developers and data science teams: most computer scientists and engineers are white men from Western countries age 20 to 40, and their personal experience, opinions, and indeed prejudices might feed into the process, potentially negatively affecting people who do not fit this description, like women, disabled people, elderly people, people of color, and people from developing countries.

Data might also be biased against particular groups because the bias is embedded in the specific practice or in the wider society. Consider claims that medicine uses mainly data from male patients and is therefore biased, or biases against people of color that are prevalent in wider society. If an algorithm feeds on such data, the results will also be biased. Bias in, bias out, as a 2016 *Nature* editorial put it. It has also been shown that machine learning can acquire bias from feeding on textual data from the World Wide Web since these language data reflect everyday human culture, including its biases (Caliskan, Bryson, and Narayanan 2017). Language corpora, for example, themselves contain gender biases. The concern is then that AI may perpetuate these biases, further disadvantaging historically marginalized groups. Bias may also arise if there is a correlation but no causation. To take again a criminal justice example: an algorithm may infer that if one of a defendant's parents went to prison, that defendant is more likely to be sent to prison. Even if this correlation may exist and even if the inference is predictive, it seems unfair that such a defendant would get a harsher sentence since there is no causal relation (House of Commons 2018). Finally, bias may also arise because human decision makers trust more in the accuracy of the recommendations of the algorithms than they should (CDT 2018) and disregard other information or do not sufficiently exercise their own judgment. For example, a judge may rely entirely on the algorithm and not take into account other elements. As always with AI and other automation technologies, human decisions and human interpretation play an important role, and there is always a risk of relying too much on the technology.

However, it is not clear if bias is avoidable at all or even if it should be avoided, and, if so, at what cost it should be avoided. For example, if changing the machine learning algorithm in order to decrease the risk of bias makes its predictions less accurate, should we change it? There may be a trade-off between effectiveness of the algorithm and the countering of bias. There is also the problem that if certain characteristics like race are ignored or removed, machine learning systems may identify so-called proxies for such characteristics, which also leads to bias. For example, in the case of race it could be that other variables that are correlated with race such as, for example, postcode, are selected by the algorithm. And is a perfectly unbiased algorithm possible? There is no consensus among philosophers or indeed in society about what perfect justice or fairness is. Furthermore, as remarked in the previous chapter, the data sets used by the algorithms are abstractions from reality and the result of human choices, and are hence never neutral (Kelleher and Tierney 2018). Bias permeates our world and societies; thus, although a lot can and should be done to minimize bias, AI models will never be entirely free from bias (Digital Europe 2018).

Moreover, surely algorithms used for decision making are *always* biased in the sense of being discriminatory: they are meant to discriminate between various possibilities. For example, in a recruiting process the screening of CVs is meant to be biased and discriminatory toward those characteristics of the candidate that best fit the position. The ethical and political question is if a particular discrimination is unjust and unfair. But, again: views on justice and fairness differ. This renders the bias issue not only technical but also connects it to political discussions about justice and fairness. For example, it is controversial whether positive discrimination or affirmative action, which tries to undo bias by creating a positive bias toward the disadvantaged individuals or groups, is just. Should justice be blind and impartial—and hence should algorithms be blind to race, for example—or does justice mean creating an advantage for those who are already disadvantaged, thus amounting to a (corrective kind of) partiality and discrimination? And should policy in a democratic context prioritize the protection of the interests of the majority or focus on promoting the interests of a minority, albeit a historically or currently disadvantaged minority?

This brings us to the question about measures. Even if we agree there is bias, there are various ways of dealing with the problem. This includes technological ways but also societal and political measures and education. Which measures we should take is controversial, and depends again on our notion of justice and fairness. For example, the question regarding affirmative action raises the more general issue of whether we should accept the world as it is or actively shape the future world in a way that avoids

Should justice be blind and impartial—or does justice mean creating an advantage for those who are already disadvantaged?

perpetuating injustices of the past. Some argue that one should use a data set that mirrors the real world. The data may represent prejudices in society and the algorithm may model existing biases people have, but this is not a problem developers should be worried about. Others argue that such a data set exists only because centuries of bias, that this bias and discrimination is unjust and unfair, and that therefore one should change that data set or the algorithm in order to promote affirmative action. For example, in response to results from the Google search algorithm that seem biased against female math professors, one could say that this simply mirrors the world as it is (and that this mirroring is exactly what the search algorithm should do); or we could have the algorithm prioritize images of female math professors in order to change perception and perhaps change the world (Fry 2018). One could also try to establish developer teams that are more diverse in terms of background, opinion, and experience, and that better represent the groups potentially affected by the algorithm (House of Commons 2018).

The mirror view does not work if the training data do not mirror the world as it is and contain old data that do not reflect the current situation. Decisions based on these data then help to perpetuate the discriminatory past rather than prepare for the future. Moreover, another objection against the mirror view is that even if a model mirrors the world as it is, this can lead to discriminatory action and other harms with regard to specific individuals and groups. For example, based on profiles made by an AI, credit firms may deny loans to applicants on the basis of where they live, or online sites may charge some customers more than others based on customer profiles created by AIs. Profiles may also follow individuals across domains (Kelleher and Tierney 2018). And a seemingly simple autocomplete function may falsely link your name to crime (which may lead to terrible consequences), even if the search AI behind it correctly mirrors the world in the sense that most people intend to search for the name of the criminal instead of your name. Another, perhaps less obvious example of bias: a music retrieval system used by services such as Spotify, which makes recommendations on the basis of current behavior (the music tracks people click on), may discriminate against music and musicians that are less mainstream. Even if it mirrors the world as it is, this leads to a situation in which some musicians cannot make a living from their music and to communities that feel unrecognized and not respected.

Again, while these are clear cases of problematic discrimination, one should always ask: is discrimination in a particular case just or not? And if it is deemed unjust, what can be done about it and by whom? For example, what can computer scientists do about it? Should they make the training data sets more diverse, perhaps even creating "idealized" data and data sets as Eric Horvitz (Microsoft) has suggested (Surur 2017)? Or should the data sets mirror the world? Should developers build in positive discrimination in their algorithms, or create "blind" algorithms? How to deal with bias in AI is not a merely technical question; it is a political and philosophical one. The question is what kind of society and world we want, whether we should try to change it, and if so, what ways of changing it are acceptable and fair. It is also a question that is as much about humans as it is about machines: do we think *human* decision making is just and fair, and if not, what is the role of AI? Perhaps AI could teach us something about humans and human societies by revealing our biases. And discussing AI ethics may reveal social and institutional power imbalances.

Discussions about AI ethics thus reach deep into sensitive societal and political issues, which are related to normative philosophical questions about, for example, justice and fairness and into philosophical and scientific questions about humans and human societies. One of these issues is the future of work.

The Future of Work and the Meaning of Life

Automation powered by AI is predicted to radically transform our economies and societies, raising questions about not only the future and meaning of work but also the future and meaning of human life.

First there is the worry that AI will destroy jobs, perhaps causing massive unemployment. There is also the question what kind of jobs will be taken over by AI: only so-called blue-collar jobs, or also others? A famous report by Benedikt Frey and Michael Osborne (2013) predicts that 47 percent of all jobs in the United States could be automated. Other reports have less dramatic figures, but most predict that job loss will be significant. Many authors agree the economy has been and will continue to be greatly transformed (Brynjolfsson and McAffee 2014), including substantial effects on employment now and in the future. And job loss due to AI is predicted to hit all kinds of workers, not only blue-collar ones, since AI is increasingly able to do complex cognitive tasks. If this is true, how can we prepare the new generations for this future? What should they learn? What should they do? And what if AI benefits some people more than others?

With this last question, we touch again on issues of justice and fairness, which have occupied thinkers in political philosophy for ages. If AI would create a wider gap between rich and poor, for example, is this just? And if not, what can be done about it? One could also frame the problem in terms of inequality (will AI increase inequality in societies and in the world?) or in terms of vulnerability: will the employed, wealthy, and educated in

Automation powered
by AI is predicted to
radically transform our
economies and societies,
raising questions about
not only the future and
meaning of work but
also the future and
meaning of human life.

technologically advanced countries enjoy the benefits of AI while the unemployed, poor, and less educated in developing countries will be far more vulnerable to the negative impacts (Jansen et al. 2018)? And to take up a more recent ethical and political concern: What about environmental justice? What is the impact of AI on the environment and our relation to the environment? What does "sustainable AI" mean? There is also the question whether AI ethics and politics should be only about the values and interests of humans or not. (See chapter 12.)

Another rather existential question concerns the meaning of work and human lives. The worry about job destruction assumes that work is the only value and the only source of income and meaning. But if jobs are the only thing of value, then we should probably create more mental illnesses, smoke more, and get more obese, since these problems tend to create jobs.[1] We don't want that. Clearly, we think that other values are more important than job creation in itself. And why rely on jobs for income and meaning? We could organize our societies and economies in a different way. We could decouple work and income, or rather what we consider "work" and income. Today many people do unpaid work, for example in the household and in care for children and elderly. Why is this not "work"? Why would it be less meaningful to do that kind of work? And why do we not make it a source of income? Moreover, some people think that automation could give us more of what is now called leisure. Maybe we can do more pleasurable and creative things, not necessarily in the form of a job. We can, in other words, question the idea that a meaningful life is only a life spent doing paid work that is prestructured by others or that takes place within a so-called self-employed frame. Maybe we can enforce measures such as "basic income" in order to allow everyone to do what they think is meaningful. Thus, in response to the future of work problem, we can think about what makes work meaningful, what kind of work humans should (be allowed to) do, and how we can reorganize our societies and economies in such a way that income is not limited to jobs and employment.

That being said, so far utopian ideas about leisure societies and other postindustrial paradises have not been realized. We have already had several waves of automation from the nineteenth century until now, but to what extent have the machines liberated and emancipated us? Perhaps they have taken over some of the dirty and dangerous work, but they have also been used for exploitation and have not radically changed the hierarchical structure of society. Some have benefited enormously from automation, whereas others have not. Perhaps fantasies about not having jobs are a luxury reserved only for those on the winning side. And have the machines freed us up to have more meaningful lives? Or do they threaten the very possibility of such lives? This is a long-standing discussion and there are no easy answers to these questions, but the concerns raised are good reasons to at least be skeptical about the brave new world painted by the AI prophets.

Furthermore, perhaps work is not necessarily toil that needs to be avoided or exploitation that needs to be resisted; a different view is that work has value, that it gives the worker purpose and meaning, and that it has various benefits such as social connections with others, belonging to something larger, health, and opportunities to exercise responsibility (Boddington 2016). If this is the case, then perhaps we should *reserve* work for humans—or at least some kinds of work, meaningful work that offers opportunities for these goods to be realized. Or at least some *tasks*. AI need not take over entire jobs, but it could take over some less meaningful tasks. We can collaborate with AIs. For example, we could choose not to delegate creative work to AIs (something Bostrom proposes) or we could choose to collaborate with AIs to do creative things. The concern here may be that if machines take over everything we do now in life, there would be nothing left for us to do and we would find our lives meaningless. However, this is a big "if": keeping in mind the skepticism about what AI can do (see chapter 3) and the fact that so many of our activities are not "work" but are very meaningful, we will probably have plenty left to do. The question is then not what humans will do when *all* their work and activities are done by machines, but rather which tasks we want to or have to reserve for humans and what the roles of AI could be, if any, in supporting us in these tasks in ways that are ethically good and societally acceptable.

To conclude, AI ethics makes us think about what a good and fair society is, what a meaningful human life is, and what the role of technology is and could be in relation to these. Philosophy, including ancient philosophy, may well be a source of inspiration for thinking about today's technologies and their potential and actual ethical and societal problems. If AI raises these ancient questions about the good and meaningful life once again, we have resources in various philosophical and religious traditions that can help us in addressing these questions. For example, as Shannon Vallor (2016) has argued, the tradition of virtue ethics developed by Aristotle, Confucius, and other ancient thinkers may still help us today to think about what human flourishing is and should be in a technological age. In other words, it could be that we have already answers to these questions, but we need to do some work to think about what the good life means in the context of today's technologies, including AI.

However, the idea of developing "an AI ethics of the good life" and an AI ethics for the real world in general face a number of problems. The first is *speed*. The model of virtue ethics Western philosophy has inherited from Aristotle assumes a slowly changing society in which technology does not change so quickly and in which people have time to learn practical wisdom; it is not clear how it can be used to cope with a fast-changing society (Boddington 2016) and the rapid development of technologies such as AI. Do we still have time to respond and to develop and communicate practical wisdom in relation to the use of technologies such as AI? Does ethics come too late? When philosophy's owl of Minerva finally spreads its wings, the world may already have been altered beyond recognition. What is and should be the role of such an ethics in the context of real-world developments?

Second, given the diversity and plurality of views on this within societies and cultural differences between societies, questions about the good and meaningful life with technology may well be answered differently in different places and contexts, and in practice they will be subject to all kinds of political processes that may or may not end in consensus. Acknowledging this diversity and plurality might lead to a more pluralist approach. It also might take the form of relativism. Twentieth-century philosophy and society theory, especially so-called postmodernism, have raised much skepticism with regard to answers that present themselves as universal while having emerged from a particular geographical, historical, and cultural context (e.g., "the West") and in relation to particular interests and power relations. It has also been questioned if politics should aim at consensus (see Chantal Mouffe's work, e.g., Mouffe 2013); is consensus always desirable, or could agonistic struggle over the future of AI also have some benefits? Moreover, there is also a problem concerning *power*: thinking about ethics in the real world means thinking not only about *what* needs to be done with regard to AI but also about *who* will and should decide about the future of AI and hence the future of our society. Consider again the issues of totalitarianism and the power of large corporations. If we reject totalitarianism and plutocracy, what does democratic decision making with regard to AI mean? What kind of knowledge about AI is needed on the part of politicians and citizens? If there is too little understanding of AI and its potential problems, we face a danger of technocracy or simply no AI policy at all.

Yet, as the next chapter shows, at least one of the AI-relevant political processes that has emerged recently displays the ambition to be timely. It is also proactive, aims at consensus, shows a surprising degree of convergence, seems to adhere to a kind of unashamed universalism, is based on expert knowledge, and pays at least lip service to the ideals of democracy, serving the public good and public interest, and involving all stakeholders: AI policymaking.

POLICY PROPOSALS

**What Needs to Be Done and Other Questions
Policymakers Have to Answer**

Given the ethical problems with AI, it is clear that something should be done. Most AI policy initiatives therefore include ethics of AI. Today there are a lot of initiatives in this area and this should be applauded. However, it is not so clear *what* should be done, what precise course of action should be taken. For example, it is not so clear how to deal with transparency or bias, given the technologies as they are, existing bias in society, and divergent views on justice and fairness. There are also many possible measures to choose from: policy can mean regulation by means of laws and directives, say, legal regulation, but there are also other strategies that may or may not be connected to legal regulation, such as technological measures, codes of ethics, and education. And within regulation there are not only laws but also standards such as ISO norms. Moreover, other sorts of questions also need to be answered in policy proposals: not only *what* should be done, but also *why* it should be done, *when* it should be done, *how much* should be done, *by whom* it should be done, and what the *nature, extent,* and *urgency of the problem* are.

First, it is important to justify the measures proposed. For example, a proposal may draw on principles of human rights to justify a proposal to reduce biased algorithmic decision making. Second, in response to technology development, policy often comes too late, when the technology is already embedded in society. Instead, one can try to make policy before the technology is fully developed and used. For AI this is still possible, to some extent, although a lot of AI is already out there. The time dimension is also relevant with regard to the temporal scope of the policy: is it only meant for the next five or ten years, or is it meant as a framework for the longer term? Here we need to make choices. For example, one could disregard long-term predictions and focus on the near future, as most proposals do, or one could offer a vision of the future of humanity. Third, not everyone agrees that solving the problems requires a lot of new measures. Some people and organizations have argued that current legislation is enough to deal with AI. If this is the case, then it seems that not much needs to be done by lawmakers, but more needs to be done by those who interpret law and those who develop AI. Others think that we need to fundamentally rethink society and its institutions, including our legal systems, in order to deal with the underlying problems and to prepare future generations. Fourth, a policy proposal should be clear about who should take action. This may be not one actor but more than one. As we have seen, many hands are involved in any technological action. This raises the question of how to distribute responsibility for policy and change: is it mainly up to governments to take action, or should, for example, businesses and industry develop their own course of action to ensure ethical AI? When it comes to business, should one address only large corporations or also small and medium-sized businesses? And what is the role of individual (computer) scientists and engineers? What is the role of citizens?

Fifth, answering what should be done, how much should be done, and other questions crucially depends on how one defines the nature, extent, and urgency of the problem itself. For example, there is a tendency in technology policy (and indeed in AI ethics) to see new problems everywhere. However, in the previous chapter we have seen that many problems may not be unique to a new technology, but perhaps existed long before. Furthermore, as the discussion about bias has also shown, what we propose to do depends on how we define the problem: is it a problem of justice, and if so, what kind of justice is threatened?

The definition will shape the measures one proposes. For example, if one proposes measures of affirmative action, then this is rooted in a particular problem definition. Finally, also playing a role is the very definition of AI, which is always contestable and which matters for the scope of the policy. For example, is it possible and desirable to distinguish clearly between AI and smart autonomous algorithms, or between AI and automation technologies? All these questions render AI policymaking a potentially controversial business. And indeed, we do find many disagreements and tensions, for example on how much new legislation is needed, on which principles exactly to draw for justifying one's measures, and on the question whether ethics should be balanced with other considerations (e.g., competitiveness of businesses and the economy). However, if we consider the actual policy documents, we also find a remarkable degree of convergence.

Ethical Principles and Justifications

The widely shared intuition that there is an urgency and importance in dealing with the ethical and societal challenges raised by AI has led to an avalanche of initiatives and policy documents that not only identify some ethical problems with AI but also aim to provide normative guidance for policy. AI policy with an ethical component

The widely shared
intuition that there is an
urgency and importance
in dealing with the
ethical and societal
challenges raised by AI
has led to an avalanche
of initiatives and policy
documents.

has been proposed by a range of actors, including governments and governmental bodies such as national ethics committees, tech companies such as Google, engineers and their professional organizations such as IEEE, intergovernmental organizations such as the EU, nongovernmental nonprofit actors, and researchers.

If we review some recent initiatives and proposals, it turns out that most documents start with the justification of policy by articulating principles and then make some recommendations with regard to the ethical problems identified. As we will see, these *problems and principles* are very similar. Initiatives often rely on general ethical principles and principles from professional ethical codes. Let me review some proposals.

Most proposals reject the science fiction scenario in which superintelligent machines take over. For example, under the presidency of Obama, the US government published the report "Preparing for the Future of Artificial Intelligence," which explicitly claims that the long-term concerns about superintelligent general AI "should have little impact on current policy" (Executive Office of the President 2016, 8). Instead, the report discusses current and near future problems raised by machine learning, such as bias and the problem that even developers may not understand their system well enough to prevent such outcomes. The report emphasizes that AI is good for innovation and economic growth and stresses self-regulation, but says that the US government can monitor the safety and fairness of applications and if necessary adapt regulatory frameworks.

Many countries in Europe have AI strategies now that include an ethical component. "Explainable AI" is a goal shared by many policymakers. The UK's House of Commons (2018) says that transparency and the right to explanation is key for algorithmic accountability, and that industries and regulators should tackle biased algorithmic decision making. The UK's House of Lords Select Committee on AI also examines the ethical implications of AI. In France, the Villani report proposes working toward "meaningful AI" that does not reinforce problems of exclusion, increase inequality, or lead to a society in which we are governed by black box algorithms: AI should be explainable and environmentally friendly (Villani 2018). Austria has recently set up a national advisory council dedicated to robotics and AI[1] that has made policy recommendations based on human rights, justice and fairness, inclusiveness and solidarity, democracy and participation, nondiscrimination, responsibility, and similar values. Its white paper also recommends explainable AI and explicitly says that responsibility remains with humans; AIs cannot be morally responsible (ACRAI 2018). International organizations and conferences are also very active. For example, the International Conference of Data Protection and Privacy Commissioners has published a declaration on ethics and data protection in AI, including principles of fairness, accountability, transparency and intelligibility, responsible design and privacy by design (a concept that calls for taking privacy into account throughout the entire engineering process), empowerment of individuals, and the reduction and mitigation of biases or discrimination (ICDPPC 2018).

Some policymakers frame their aim in terms of "trustworthy AI." The European Commission, for example, undoubtedly one of the main global players in the area of AI policymaking, puts a lot of weight on the term. In April 2018, it set up a new High-Level Expert Group on Artificial Intelligence to create a new set of AI guidelines; in December 2018 the group released a draft working document with ethics guidelines that calls for a human-centric approach to AI and the development of trustworthy AI, which respects fundamental rights and ethical principles. The rights mentioned are human dignity, freedom of the individual, respect for democracy, justice and the rule of law, and citizens' rights. The ethical principles are beneficence (do good) and no harm, autonomy (preserve human agency), justice (be fair), and explicability (operate transparently). These principles are familiar from bioethics, but the document adds explicability and includes interpretations that highlight the specific ethical problems raised by AI. For example, the no harm principle is interpreted as requiring that AI algorithms must avoid discrimination, manipulation, and negative profiling, and must protect vulnerable groups such as children and immigrants. The justice principle is interpreted as including the demand that developers and implementers of AI need to ensure that individuals and minority groups maintain freedom from bias. The principle of explicability is seen as requiring that AI systems be auditable and "comprehensible and intelligible by human beings at varying levels of comprehension and expertise" (European Commission AI HLEG 2018, 10). The final version, released in April 2019, specifies that explainability is not only about explaining the technical process but also about the related human decisions (European Commission AI HLEG 2019, 18).

Previously another EU advisory body, the European Group on Ethics in Science and New Technologies (EGE), released a statement on AI, robotics, and autonomous systems, proposing the principles of human dignity, autonomy, responsibility, justice, equity, solidarity, democracy, rule of law and accountability, security and safety, data protection and privacy, and sustainability. The principle of human dignity is said to imply that people have to be made aware whether they are interacting with a machine or with another human being (EGE 2018). Note also that the EU already has existing regulations in place relevant to the development and use of AI. The General Data Protection Regulation (GDPR), which was enacted in May 2018, aims to protect and empower all EU citizens with regard to data privacy. It includes principles such as the right to be forgotten (the data subject can ask to erase his or her personal data and halt the further processing of the data) and privacy by design. It also gives data subjects the right to access "meaningful information about the logic involved" in automated decision making and information about the "envisaged consequences" of such processing (European Parliament and the Council of the EU 2016). The difference with the policy documents is that here these principles are legal requirements. It is legislation that is enforced: organizations in breach of GDPR can be fined. However, it has been questioned whether the provisions of the GDPR amount to a full right of explanation of the decision (Digital Europe 2018) and, more generally, whether it offers enough protection against the risks of automated decision making (Wachter, Mittelstadt, and Floridi 2017). The GDPR provides a right to be informed about automated decision making but does not seem to demand an explanation about the rationale for any individual decision. This is also a concern when it comes to decision making in the legal sphere. A Council of Europe study, based on work by a committee of human rights experts, has demanded that individuals have the right to a fair trial and due process in terms comprehensible to them (Yeung 2018).

Legal discussions are of course highly relevant for discussions about AI ethics and AI policy. Turner (2019) has discussed comparisons with animals (how they are and have been treated by the law and whether they have rights) and has reviewed a number of legal instruments with regard to what they could mean for AI. For example, when harm has been inflicted, the question of negligence concerns whether a person was under a duty of care to prevent harm, even if said harm was not intended. This could be applied to the designer or trainer of the AI. But how easy is it to foresee the consequences of AI? Criminal law, by contrast, requires the intention to do harm. But this is often not the case with AI. Product liability, on the other hand, does not concern the fault of individuals but has the company who produced the technology pay for damages, regardless of fault. This could be one possible solution to legal responsibility for AI. Intellectual property laws are also relevant to AI, such as copyright and patents, and discussions have emerged about "legal personality" for AI, a legal fiction but an instrument that is currently applied to companies and various organizations. Should it also be applied to AI? In a controversial resolution of 2017, the European Parliament suggested that giving the most sophisticated autonomous robots the status of electronic persons is one possible legal solution to the issue of legal responsibility—an idea that has *not* been taken up by the European Commission in its 2018 strategy for AI.[2] Others have aggressively opposed the very idea of giving rights and personhood to machines, arguing, for instance, that it would then become difficult if not impossible to hold anyone accountable since people will seek to exploit the concept for selfish ends (Bryson, Diamantis, and Grant 2017). There was also the famous case of Sophia, a robot granted "citizenship" by Saudi Arabia in 2017. Such a case raises again the question regarding moral status of robots and AIs (see chapter 4).

AI policy has also been proposed beyond North America and Europe. China, for instance, has a national AI strategy. Its development plan recognizes that AI is a disruptive technology that can affect social stability, impact law and social ethics, violate personal privacy, and create safety risks; the plan therefore recommends to strengthen forward-looking prevention and minimizing risk (State Council of China 2017). Some actors in the West tell a competition narrative: they fear that China will overtake us or even that we are approaching a new world war. Others try to *learn* from China's strategy. Researchers may also ask how different cultures deal with AI in different ways. AI research itself can contribute to taking a more cross-cultural or comparative perspective on AI ethics, for example, when it reminds us of differences between individualist and collectivistic cultures when it comes to moral dilemmas (Awad et al. 2018). This could raise problems for AI ethics if it aims to be universal. One could also explore how narratives about AI in China or Japan, for example, differ from Western narratives. Yet in spite of cultural differences, it turns out that AI ethics policies are remarkably similar. While China's plan places more emphasis on social stability and the collective good, the ethical risks identified and the principles mentioned are not that different from those proposed by Western countries.

But, as mentioned before, AI ethics policy is also not at all restricted to governments and their committees and bodies. Academics have also taken initiatives. For example, the Montreal Declaration Responsible AI has been proposed by the University of Montreal and involved consultation of citizens, experts, and other stakeholders. It says that the development of AI should promote the well-being of all sentient creatures and the autonomy of human beings, eliminate all types of discrimination, respect personal privacy, protect us from propaganda and manipulation, promote democratic debate, and make various players responsible for working against the risks of AI (Université de Montréal 2017). Other researchers have proposed the principles of beneficence, non-maleficence, autonomy, justice, and explicability (Floridi et al. 2018). Universities such as Cambridge and Stanford work on the ethics of AI, often from an applied ethics perspective. People working in professional ethics also do helpful work. For example, the Markkula Centre for Applied Ethics at Santa Clara University has offered a number of ethical theories as a toolkit for technology and engineering practice, which

In spite of cultural differences, it turns out that AI ethics policies are remarkably similar.

may also inform AI ethics.[3] And philosophers of technology have recently shown a lot of interest in AI.

We also find initiatives on AI ethics in the corporate world. For example, the Partnership on AI includes companies such as DeepMind, IBM, Intel, Amazon, Apple, Sony, and Facebook.[4] Many companies recognize the need for ethical AI. For example, Google has published ethics principles for AI: providing social benefit, avoiding creating or reinforcing unfair bias, enforcing safety, maintaining accountability, maintaining privacy design, promoting scientific excellence, and limiting potentially harmful or abusive applications such as weapons or technologies that violate principles of international law and human rights.[5] Microsoft talks about "AI for Good" and proposes the principles of fairness, reliability and safety, privacy and security, inclusiveness, transparency, and accountability.[6] Accenture has proposed universal principles of data ethics, including respecting the persons behind the data, privacy, inclusion, and transparency.[7] And although in corporate documents the emphasis tends to be on self-regulation, some companies recognize the need for external regulation. Apple's CEO Tim Cook has said that tech regulation, for example, to ensure privacy, is inevitable because the free market is not working.[8] Yet there is debate about whether this requires new regulation. Some support the path of regulation, including new laws. California has already proposed a bill that requires the disclosure of bots: it is unlawful to use a bot if it is done in a way that misleads another person about its artificial identity.[9] Others take a more conservative position. Digital Europe (2018), which represents the digital industry in Europe, has argued that today's legal framework is equipped to address concerns related to AI, including bias and discrimination, but that in order to build trust, transparency, explainability, and interpretability are important: people and businesses should understand when and how algorithms are used in decision making, and we need to provide meaningful information and facilitate the interpretation of algorithmic decisions.

Nonprofit actors also play a role. For example, the international Campaign to Stop Killer Robots raises many ethical questions regarding military applications of AI.[10] From a transhumanist side, there are the Asilomar AI principles, agreed on by academic and industry participants of a conference convened by the Future of Life Institute (Max Tegmark and others). The overall aim is to keep AI beneficial and to respect ethical principles and values such as safety, transparency, responsibility, value alignment, privacy, and human control.[11] There are also professional organizations that work on AI policy. The Institute of Electrical and Electronics Engineers (IEEE), which claims to be the world's largest technical professional organization, has put forward a Global Initiative on Ethics of Autonomous and Intelligent Systems. After discussions among experts, the Initiative has produced a document with a vision for "ethically aligned design," proposing that the design, development, and implementation of these technologies should be guided by the general principles of human rights, well-being, accountability, transparency, and awareness of misuse. Implementing ethics in global technical standards could be an effective way of contributing to the development of ethical AI.

Technological Solutions and the Question of Methods and Operationalization

The IEEE Global Initiative shows that, in terms of measures, some policy documents focus on technological solutions. For example, as mentioned in the previous chapter, some researchers have called for explainable artificial intelligence, for opening the black box. There are good reasons for wanting to do this: explaining the rationale behind one's decision is not only ethically required but also an important aspect of human intelligence (Samek, Wiegand, and Müller 2017). The idea of explainable AI or transparent AI is then that the actions and decisions made by AIs should be easily understood. As we've seen, this idea is difficult to implement in the case of machine learning that uses neural networks (Goebel et al. 2018). But policy can of course support research in this direction.

In general, it is an excellent idea to embed ethics in the design of new technologies. Ideas such as ethics by design or value-sensitive design, which have their own history,[12] can help us to create AI in a way that leads to more accountability, responsibility, and transparency. For example, ethics by design could include the requirement that traceability is ensured at all stages (Dignum et al. 2018), thus contributing to the accountability of AI. The idea of traceability can be taken literally, in the sense of recording data about the behavior of the system. Winfield and Jirotka (2017) have called for implementing an "ethical black box" in robots and autonomous systems, which records what the robot does (data from sensors and from the "internal" state of the system) in a way similar to the black box installed in airplanes. This idea could also be applied to autonomous AI: when something goes wrong, such data might help us to explain what exactly went wrong. This may assist ethical and legal analyses of the case. Moreover, as the researchers rightly remark, we can learn something from the aircraft industry, which is highly regulated and has tough safety certification processes and visible processes of accident investigation. Could similar regulatory and safety infrastructures be installed for AI? To make a comparison with another transportation field, the automobile industry has also proposed certification or a kind of "driver's license" for AI autonomous vehicles.[13] Some researchers go further and aim at creating moral machines,

Ideas such as ethics by design or value-sensitive design can help to create AI in a way that leads to more accountability, responsibility, and transparency.

attempting "machine ethics" in the sense that machines themselves can make ethical decisions. Others argue that this is a dangerous idea and that this should be reserved for humans, that it is impossible to create full ethical agents, that there is no need for machines to be full ethical agents and that it is enough for them to be safe and law abiding (Yampolskiy 2013), or that there could be forms of "functional morality" (Wallach and Allen 2009) that do not amount to full morality but still render the machine relatively ethical. This discussion, which connects again to the issue of moral status, is relevant, for example, in the case of self-driving cars: to what extent is it necessary, possible, and desirable to build ethics into these cars, and what kind of ethics should this be and how should it be technically implemented?

Policymakers tend to endorse many of these directions in AI research and innovation, such as explainable AI and, more generally, embedding ethics in design. For example, next to nontechnical methods such as regulation, standardization, education, stakeholder dialogue, and inclusive design teams, the High-Level Expert Group report mentions a number of technical *methods* including ethics and rule of law by design, architectures for trustworthy AI, testing and validating, traceability and auditability, and explanation. For example, ethics by design can include privacy by design. The report also mentions some ways in which trustworthy AI can be *operationalized*, such as traceability as a way to contribute to transparency: in the case of rule-based AI it should be clarified how the model has been built, and in the case of learning-based AI the method of training the algorithm should be clarified, including how the data were gathered and selected. This should ensure that the AI system is auditable, particularly in critical situations (European Commission AI HLEG 2019).

The question of methods and operationalization is crucial: it is one thing to name a number of ethical principles and quite another to figure out how to implement them in practice. Even concepts such as privacy by design, which are supposed to be closer to the process of development and engineering, are usually formulated in a rather abstract and general way: *it remains unclear what exactly we should do*. This takes us to the next chapter for a brief discussion of some challenges for AI ethics policy.

CHALLENGES FOR POLICYMAKERS

Proactive Ethics: Responsible Innovation and Embedding Values in Design

Perhaps unsurprisingly, AI ethics policy faces numerous challenges. We have seen that some policy proposals endorse a vision of AI ethics that is *proactive*: we need to take ethics into account at the early stage of the development of AI technology. The idea is to avoid ethical and societal problems created by AI that would be hard to deal with once they have arrived. This is in line with ideas about responsible innovation, embedding values in design, and similar ideas proposed over recent years. It shifts the problem from having to deal with the negative effects of technologies that are already widely used to taking responsibility for technologies that are being developed today.

However, it is not easy to foresee the unintended consequences of new technologies at design stage. One way to mitigate this problem is to construct scenarios about future ethical effects. There are various methods for practicing ethics in research and innovation (Reijers et al. 2018), one of which is not only to study and assess the impact of current AI narratives (Royal Society 2018) but also to create new, more concrete narratives about particular AI applications.

Practice Oriented and Bottom Up: How Can We Translate These to Practice?

Responsible innovation is not only about embedding ethics in design, but also requires taking into account the opinions and interests of various stakeholders. Inclusive governance entails broad stakeholder involvement, public debate, and early societal intervention in research and innovation (Von Schomberg 2011). This can mean, for example, organizing focus groups and using other techniques to see what people think about the technology.

This more bottom-up responsible innovation approach is somewhat in tension with the applied ethics approach of most policy documents, which are rather top down and remarkably abstract. First, policies are often created by experts, without input from a broad range of

Responsible innovation is not only about embedding ethics in design, but also requires taking into account the opinions and interests of various stakeholders.

stakeholders. Second, even if they endorse principles such as ethics by design, they tend to remain too vague about what applying these principles means in practice. To make AI policy work, it remains a huge challenge to build a bridge between, on the one hand, abstract, high-level ethical and legal principles and, on the other hand, the practices of technology development and use in particular contexts, the technologies, and the voices of those who are part of these practices and work in these contexts. This bridging work is left to the addressees of the proposals. Can and should more be done, at the earlier stage of policymaking? At the very least, more work on the "how" is required alongside the "what": the methods, procedures, and institutions we need for making AI ethics work in practice. We need to pay more attention to *process*.

With regard to the "who" question concerning AI ethics, we need more room for *bottom*-up next to top-down, in the sense of listening more to researchers and professionals who work with AI in practice and indeed to people potentially disadvantaged by AI. If we endorse the ideal of democracy and if that concept includes inclusiveness and participation in decision making about the future of our societies, then hearing the voice of stakeholders is not optional but ethically and politically required. While some policymakers engage in some form of stakeholder consultation (for example, the European Commission has its AI Alliance),[1] it remains questionable if such efforts really reach the developers, the end users of the technology, and—most importantly—those who will have to carry most of the risks and live with its negative consequences. How democratic and participative is the decision making and policy about AI, really?

The ideal of democracy is also endangered by the fact that power is concentrated in the hands of a relatively small number of large corporations. Paul Nemitz (2018) has argued that such accumulation of digital power in the hands of a few is problematic: if a handful of companies exercise power not only over individuals—by profiling us, they centralize power—but also over infrastructures for democracy, then in spite of their best intentions to contribute to ethical AI, such companies also put up barriers to it. It is then necessary to regulate and set boundaries to safeguard public interest, and to make sure that these companies do not shape the rules by themselves. Murrah Shanahan has also pointed to the "self-perpetuating tendency for power, wealth, and resources to concentrate in the hands of a few" (2015, 166), which makes it difficult to bring about a more equitable society. It also renders people vulnerable to all kinds of risks, including exploitation and violations of privacy, for example what a Council of Europe study calls "the chilling effect of data repurposing" (Yeung 2018, 33).

If we compare the situation with environmental policy, we may also be pessimistic about the possibility that countries will take effective and collaborative action concerning AI ethics. Consider, for example, the political processes around climate change in the United States, where sometimes even the very *problem* of global warming and climate change is denied and where powerful political forces work against taking action, or the rather limited success of international climate conferences to agree on a common and effective climate policy. Those seeking global action in light of the ethical and societal problems raised by AI may face similar difficulties. Often interests other than public good prevail, and there is far too little intergovernmental policy on new digital technologies, including AI. One exception is the global interest in banning automatic lethal weapons, which also have an AI aspect to it. But this remains an exception, and is also not supported by all countries (it remains controversial in the US, for example).

Moreover, albeit well-intended, ethics by design and responsible innovation have their own limitations. First, methods such as value-sensitive design presuppose that we can articulate our values, and efforts to build moral machines assume that we can fully articulate our ethics. But this is not necessarily the case; our everyday ethics may not be a matter of fully articulate reasoning at all. Sometimes we respond to ethical problems without being able to fully justify our response (Boddington 2017). To borrow a term from Wittgenstein: our ethics is not only embodied but also embedded in a *form of life*. It is deeply connected to the way we do things as embodied and social beings, and as societies and cultures. This sets limits to the project of fully articulating ethics and moral reasoning. It poses a problem for the project of developing moral machines and challenges the assumptions that ethics and democracy can be *fully* deliberative. It is also a problem for those policymakers who think that AI ethics can be dealt with fully by means of a list of principles or by means of specific legal and technical methods. We absolutely need methods, procedures, and operations. But these are not enough; ethics does not work like a machine, and neither do policy and responsible innovation.

Second, these approaches can also be a *barrier* to ethics when it would be ethically required to stop the development of the technology. Often they function in practice as a kind of oil that helps to lubricate the machinery of innovation, to enhance profit making, and to ensure the acceptability of the technology. This may not be necessarily bad. But what if the ethical principles imply that the technology, or a particular application of the technology, should actually be halted or paused? Crawford and Calo (2016) have argued that value-sensitive design and responsible innovation tools work on the assumption that the technology will be developed; they are less helpful when it comes to deciding whether it should be built at all. For example, in the case of advanced AI such as new machine learning applications, it might be that the technology is still unreliable or has serious ethical drawbacks, and that at least some applications should not be used (yet). Whether or not stopping is always the best solution, the point is that we should at least have the *space to ask the question* and to decide. If this critical space is lacking, responsible innovation remains a fig leaf for doing business as usual.

Toward a Positive Ethics

That being said, generally speaking AI ethics is not necessarily about banning things (Boddington 2017). Another barrier to getting AI ethics to work in practice is that many actors in the AI field such as companies and technical researchers still think of ethics as a constraint, as something negative. This idea is not totally misguided: often ethics has to constrain, has to limit, has to say that something is unacceptable. And if we take AI ethics seriously and implement its recommendations, we might face some trade-offs, in particular in the short term. Ethics may have a cost: in terms of money, time, and energy. However, by reducing risks, ethics and responsible innovation support the long-term, sustainable development of businesses and of society. It is still a challenge to convince all the actors in the AI field, including policymakers, that this is indeed the case.

AI ethics is not necessarily about banning things; we also need a *positive* ethics: to develop a vision of the good life and the good society.

Note also that policy and regulation are not only about banning things or making things more difficult; they can also be supportive, offering incentives, for example.

Furthermore, next to a negative ethics that sets limits, we also need to make explicit and elaborate a *positive* ethics: to develop a vision of the good life and the good society. While some of the ethical principles proposed above hint at such a vision, it remains a challenge to move the discussion in that direction. As previously argued, the ethical questions regarding AI are not just about technology; they are about human lives and human flourishing, about the future of society, and perhaps also about nonhumans, the environment, and the future of the planet (see the next chapter). Discussions about AI ethics and AI policy then lead us once more to the big questions we need to ask ourselves—as individuals, as societies, and perhaps as humanity. Philosophers can help our thinking about these questions. For policymakers, the challenge is to develop a broad vision of the technological future that includes ideas about what is important, meaningful, and valuable. While in general liberal democracies are set up to leave such questions to individuals and are supposed to be "thin" about matters such as the good life (a political innovation that has stopped at least some kinds of wars and has contributed to stability and prosperity), in light of the ethical and political challenges we face it would be irresponsible to neglect the more substantive, "thick" ethical questions altogether. Policy, including AI policy, should also be about positive ethics.

The way to do this for policymakers, however, is not by flying solo and taking the role of the Platonic philosopher-king, but to find the right balance between technocracy and participative democracy. The questions at hand are questions that concern us all; we all have a stake in them. Therefore, they cannot be left in the hands of the few, whether these are people in government or in large corporations. This leads us back to questions about how to make responsible innovation and participation in AI policy work. The problem is not only about power; it is also about the good: the good for individuals and the good for society. Our current ideas about the good life and the good society, if we can articulate them at all, may well need a lot more critical discussion. Let me suggest that for the West it could be helpful to at least explore the option of trying to learn from other, non-Western political systems and other political cultures. An effective and well-justified AI policy should not avoid tapping into these kinds of ethical-philosophical and political-philosophical discussions.

Interdisciplinarity and Transdisciplinarity

There are further barriers that we need to overcome if we want to make AI ethics more effective and support the responsible development of the technology, avoiding what technical researchers call a new AI "winter": the slowing down of AI development and investment. One is the lack of sufficient *interdisciplinarity* and *transdisciplinarity*. We still face a significant gap in background and understanding between, on the one hand, people from the humanities and social sciences, and, on the other hand, people from the natural and engineering sciences, both within and outside academia. So far, institutional support for a significant and substantial bridging between these two "worlds" is lacking, in academia and in wider society. But if we really want to have ethical high tech such as ethical AI, we need to bring these people and these worlds closer together, sooner rather than later.

This requires a change in how research and development is done—it should involve not only technical and business people but also people from the humanities, for example—but also a change in how we *educate* people, both young people and not so young people. We need to ensure that, on the one hand, people with a humanities background become aware of the importance of thinking about new technologies such as AI and can acquire some knowledge of these technologies and what they do. On the other hand, scientists and engineers need to be made more sensitive to the ethical and societal aspects of technology development and use. When they learn to use AI and later contribute to the development of new AI technology, ethics should be seen not as a marginal topic that has little to do with their technological practice but as constituting *an essential part of it*. Ideally, what it means to "do AI" or to "do data science" would then simply include ethics. More generally, we could consider the idea of a more diverse and holistic kind of *Bildung* or narrative that is more radically interdisciplinary and pluralistic with regard to methods and approaches, its topics, and also its media and technologies. To put it bluntly: if engineers learn to do things with texts and humanities people learn to do things with computers, there is more hope for a technology ethics and policy that works in practice.

The Risk of an AI Winter and the Danger of the Mindless Use of AI

If these directions in policy and education do not get off the ground and, more generally, if the project of ethical AI fails, we face not only the risk of an "AI winter"; the ultimate and arguably more important risk is ethical, social, and economic disaster and its related human, non-human, and environmental costs. This has nothing to do with the singularity, terminators, or other apocalyptic scenarios about the distant future, but with the slow but certain increase in the accumulation of technological risk and the resulting growth of human, social, economic, and environmental vulnerabilities. This increase in risks and vulnerabilities is related to the ethical problems indicated here and in the previous chapters, including the ignorant and reckless use of advanced automation technologies such as AI. The gap in education is perhaps exacerbating what AI risks do in general: even if it does not always directly cause new risks, it also and especially *multiplies existing risks*. So far there is no such thing as a "driver's license" for using AI, and there is no compulsory AI ethics education for technical researchers, business people, government administrators, and other people involved in AI innovation, use, and policy. There is a lot of untamed AI out there in the hands of people who don't know the risks and ethical problems, or who may have the wrong kind of expectations about the technology. The danger is, once again, the exercise of power without knowledge and (therefore) without responsibility—and, worse, others being subjected to this. If there exists such a thing as evil at all, it lives where the twentieth-century philosopher Hannah Arendt located it: in the mindlessness of banal everyday work and decisions. To assume that AI is neutral and to use it without understanding what one is doing contributes to such mindlessness and, ultimately, to the ethical corruption of the world. Education policy can help to mitigate this and thus contribute to good and meaningful AI.

A number of nagging, perhaps slightly painful questions remain, however, which are often neglected in discussions about AI ethics and policy but that at least deserve mention, if not a lot more analysis. Is AI ethics all about the good for, and value of, humans, or should we also take into account nonhuman values, goods, and interests? And even if AI ethics is mainly about humans, could it be that the question regarding AI ethics is not the most important problem for humanity to address? This question brings us to the last chapter.

IT'S THE CLIMATE, STUPID! ON PRIORITIES, THE ANTHROPOCENE, AND ELON MUSK'S CAR IN SPACE

Should AI Ethics Be Human-Centric?

While many writings on AI ethics and policy mention the environment or sustainable development, they emphasize human values and are often explicitly human-centric. For example, the HLEG's Ethics Guidelines say that a human-centric approach to AI is needed "in which the human being enjoys a unique and inalienable moral status of primacy in the civil, political, economic and social fields" (European Commission AI HLEG 2019, 10) and universities such as Stanford and MIT have framed their research policies in terms of human-centered AI.[1]

Usually this human-centricity is defined in relation to technology: the idea is that the good and dignity of humans take priority over whatever technology may require or do. The point is that technology should benefit humans and should serve humans rather than the other way around. Yet, as we have seen in the first chapters, the appropriateness of this focus on humans in AI ethics is not as obvious as it may at first seem, especially if we consider posthumanist approaches or critically question competition narratives (humans versus technology). Philosophy of technology shows that there are more—and subtler and more sophisticated—ways of defining the relation between humans and technology. Furthermore, a human-centric approach is at least nonobvious, if not problematic, in light of philosophical discussions about the environment and other living beings. In environmental philosophy and ethics there is a long-standing discussion about the value of nonhumans, especially living beings, about how to respect that value and these beings, and about the potential tensions that may arise with respecting the value of humans. For AI ethics, this implies that we should at least ask the question regarding the impact of AI on other living beings and consider the problem that there might be tension between human and nonhuman values and interests.

Getting Our Priorities Right

It could also be argued that there are other more serious problems than those caused by AI, and that it is important

A human-centric approach is at least nonobvious, if not problematic, in light of philosophical discussions about the environment and other living beings.

to get one's priorities right. This objection could arise from a consideration of global problems such as climate change, according to some *the* problem humanity needs to address and prioritize given its urgency and potential impact on the planet as a whole.

Looking at the United Nations' 2015 agenda for sustainable development (the so-called Sustainable Development Goals)[2] and its overview of global issues concerning what UN Secretary-General Ban Ki-moon called "people and planet," we see many global issues that demand ethical and political attention: rising inequalities within and among countries, war and violent extremism, poverty and malnutrition, lack of access to fresh water, lack of effective and democratic institutions, ageing populations, infectious and epidemic diseases, risks related to nuclear energy, lack of opportunities for children and young people, gender inequality and various forms of discrimination and exclusion, humanitarian crises and all kinds of human rights violations, problems related to migration and refugees, and climate change and environmental problems—sometimes related to climate change—such as more frequent and intense natural disasters and forms of environmental degradation such as drought and loss of biodiversity. In light of these massive problems, should AI be our top priority? Does AI distract from more important issues?

One the one hand, a focus on AI and other problems with technology seems out of place when so many humans are suffering and the world is plagued by so many other problems. While people in one part of the world struggle to gain access to fresh water or to survive in violent environments, people in another part of the world worry about their privacy on the internet and fantasize about a future when AIs achieve superintelligence. Ethically speaking, something fishy seems to be going on, which is related to global inequalities and injustices. Ethics and policy should not remain blind to such problems, which are not necessarily about AI at all. For example, sometimes in developing countries, *low tech* rather than high tech can help people to deal with problems since they can afford, install, and maintain it.

On the other hand, AI could cause new problems and also *aggravate existing problems* in societies and with the environment. For example, some fear that AI will widen the gap between rich and poor and that, like many digital technologies, it will increase energy consumption and create more waste. From this perspective, discussing and dealing with AI ethics is not a distraction but one of the ways we can contribute to addressing the world's problems, including environmental ones. One could thus conclude that we *also* need to pay attention to AI: yes, poverty, war, and so on are serious problems, but AI may also cause or aggravate serious problems now and in the

While people in one part of the world struggle to gain access to fresh water, people in another part of the world worry about their privacy on the internet.

future, and should be on our list of problems needing solutions. However, this does not answer the question about priorities—an important ethical and policy question. The point is not that there are easy answers to that question; the point is that the question is not even *asked* in most academic writings and policy documents on AI.

AI, Climate Change, and the Anthropocene

One of the most challenging ways to ask the question regarding priorities is to bring in the discussion about climate change and related topics such as the Anthropocene: "Why worry about AI if the urgent problem is climate change and the future of the planet is at stake?" Or to adapt a phrase from US political culture: "It's the climate, stupid!" Let me unpack this challenge and discuss its implications for thinking about AI ethics.

While some extremists reject the scientific findings, climate change is widely recognized by scientists and policymakers to be not only a serious global problem but also "one of the greatest challenges of our time," as the UN's Sustainable Development Goals text puts it. It is not only a problem in the future: global temperature and sea levels are *already* rising, which is affecting lower-lying coastal areas and countries. Soon more people will have to deal with the consequences of climate change. Many people

"Why worry about AI if the urgent problem is climate change and the future of the planet is at stake?"

conclude from this that we have to act urgently, *now*, to mitigate the risks of climate change—"mitigate" because the process may well be beyond the tipping point. The idea is that it is not only high time to do something but possibly already too late to avoid all consequences. Compared with transhumanist fears about superintelligence, this concern is much better supported by scientific evidence and has gained considerable support among the well-educated elites in the West who, understandably bored by postmodern skepticism and bureaucratized identity politics, now see a reason to focus on a problem that seems so true, so real, and so universal: climate change is *really* happening and concerns *everyone* and everything on this planet. A recent wave of activism calls attention to the climate crisis, for example Greta Thunberg's campaign and the climate strikes.

Sometimes the concept of the Anthropocene is used to frame the problem. Coined by climate researcher Paul Crutzen and biologist Eugene Stoermer, this is the idea that we are living in a geological epoch in which humanity has dramatically increased its power over the Earth and its ecosystems, turning humans into a geological force. Consider the exponential growth of human and cattle populations, increasing urbanization, the exhaustion of fossil fuels, the massive use of fresh water, the extinction of species, the release of toxic substances, and so on. Some think the Anthropocene started with the agricultural revolution; others think it took off with the industrial revolution (Crutzen 2006) or after the Second World War. In any case, a new story and new history have been created, perhaps even a new grand narrative. The concept is often used today to raise concern about global warming and climate change, and to bring together various disciplines (including the humanities) to think about the future of the planet.

Not everyone adopts the term—it is controversial even among geologists—and some have questioned its anthropocentrism. For example, Haraway (2015) has argued from a posthumanist perspective that other species and "abiotic" actors also play a role in the shifting environment. But even without a controversial concept such as the Anthropocene, climate change and (other) environmental problems are here to stay, and policy must deal with them, preferably sooner rather than later. What does this mean for AI policy?

Many researchers think that AI and big data could also help us to deal with many of the world's problems, including climate change. Like digital information and communication technologies in general, AI can contribute to sustainable development and to dealing with many environmental problems. Sustainable AI is likely to become a successful direction in research and development. However, AI could also make things worse for the environment—and hence for all of us. Consider again increased energy consumption and waste. And seen from the perspective of the Anthropocene problem, the risk is that humans could use AI to tighten their grip on the Earth, thus worsening the problem instead of solving it.

This is especially problematic if AI is seen not only as *a* solution but as the *main* solution. Consider a superintelligence scenario of an AI that knows better than us humans what is good for us: a "benign" AI that serves humanity by making humans act in their own interests and that of the planet—say, a technological equivalent of Plato's philosopher-king, a machine god. *Homo deus* (Harrari 2015) is replaced by *AI deus*, which manages our life support system for us and manages us. To solve the problems of resource distribution, for example, the AI could act as a "server," managing the access humans have to resources. Its decisions would be based on its analysis of patterns in data. This diet scenario could be combined with Promethean technological solutions such as geo-engineering. It's not just humans who need to be managed; the planet needs to be reengineered. Technology would thus be used to "fix" our problems and those of the planet.

Yet these scenarios would not only be authoritarian and violate human autonomy but would also centrally contribute to the problem of the Anthropocene itself: human hyper-agency, this time delegated by humans to machines, turns the entire planet into a resource and machine for humans. The problem of the Anthropocene is "solved" by taking it to its technocratic extreme, leading to a world of machines in which humans are first treated as children to be cared for and perhaps later become obsolete. With this kind of Big Data Anthropocene and the all-too-familiar drama of humans being replaced by machines, we land back in the scenarios of dreams and nightmares.

The New Space Craze and the Platonic Temptation

Another answer to climate change and the Anthropocene, which is also a technophile vision and is sometimes linked with transhumanist narratives, is this: we might mess up *this* planet, but we can escape from the Earth and go to space.

An iconic image of 2018 was Elon Musk's Tesla sports car floating in space.[3] Musk also has plans to colonize Mars. He is not the only dreamer: there is a growing interest in going to space. And it is more than just a dream. A lot of money is being invested in space projects. In contrast to the twentieth-century space race, it is now propelled by private companies. And not only tech millionaires but also artists are very interested in space. Musk's company SpaceX plans to send artists to orbit the moon.[4] Space tourism is another increasingly popular idea. Who wouldn't want to go to space? Space is hot.

There is nothing wrong with going to space per se. It does have potential benefits. For example, research on how to survive in more extreme environments can help us to deal with problems on Earth, experiment with sustainable technologies, and take a planetary perspective. Consider also that the problem of the Anthropocene could be formulated only because many years earlier, space technology made it possible for us to view the Earth from a distance. And considering the image of Musk's car again: some think the electric car is a solution to environmental problems, without questioning the assumption that cars are the best means of transportation and without considering how the electricity is produced. Anyway, there are interesting ideas out there.

But the space dreams *are* problematic if the result is that earthly problems are neglected, and if they are symptomatic of a tendency Hannah Arendt (1958) already diagnosed when she wrote about the human condition: too much abstraction and alienation. She suggested that science supports a desire to leave the Earth: literally, by means of space technology (in her time, Sputnik) but also by means mathematical methods that abstract and alienate from what I would call our messy earthly, embodied, and political life. From this perspective, transhumanist fantasies about superintelligence and about leaving the Earth can be interpreted as exponents of a problematic kind of alienation and escapism. It is Platonism and transhumanism writ large: the idea is to overcome not only the limitations of the human body, but also those of this other "life support system": the Earth itself. Not only the body but the Earth is seen as a prison, as something we need to escape.

One danger of AI, then, is that it enables this kind of thinking and becomes an alienation machine: an instrument to leave the Earth and deny our vulnerable, bodily, earthly, and dependent existential condition. In other words: a rocket. Again, there is nothing wrong with rockets per se. The problem is the mix of particular technologies with particular narratives. While AI can potentially be a positive force for our personal lives, for society, and for humanity, a combination of the amplification of the abstracting and alienating tendencies in science and technology with transhumanist and "trans-Earthist" fantasies may lead to a technological future that is not good for human beings and other living beings on Earth. If we escape rather than deal with our problems—for example, climate change—then we may win Mars (for now) but lose the Earth.

And as always, there is a further political side to this: some have more opportunities, money, and power to escape than others. The problem is not only that space tech and AI have real costs for the Earth and that all the money invested in space projects is not spent on real earthly problems such as war and poverty; the worry is also that the rich will be able to escape the Earth they destroy, whereas the rest of us have to stay on an increasingly unlivable planet (see, e.g., Zimmerman 2015). Like rockets and other tech, AI may become a tool for the "survival of the richest," as one commentator put it (Rushkoff 2018). Today some of this is already happening with other technologies: in cities such as Delhi and Beijing, the majority of people are plagued by air pollution whereas the rich fly out, live in less polluted areas, or buy good air by using air purification technologies. Not everyone breathes the same air. Will AI contribute to such gaps between the rich and poor, leading to more stressful and unhealthy lives for some and better lives for others? Will AI alienate us from environmental problems? It seems an ethical requirement that AI should also make life on Earth better, preferably for all of us and taking into account that for human life we depend on the Earth. Some space narratives may hinder rather than help us to reach this goal.

Return to Earth: Toward Sustainable AI

Let me return to the very practical problem of priorities and the very real and present-day risks related to climate change. What should AI ethics and policy do in light of these challenges? And when there are conflicts with the value of nonhuman lives, how can these be solved? Most people will agree that just handing over the control to an AI or escaping the Earth are not good solutions. But what is a good solution? Is there a solution? A more productive answer to these questions leads us necessarily back to philosophical questions regarding how we as humans relate to technology and to our environment. It also leads us back to the technology chapter: what can AI and data science do for us, and what can we reasonably expect from AI?

It is clear that AI can help us to tackle environmental problems. Consider climate change. AI would seem particularly suited to helping us with such complex problems. AI can help us to study the problem, for example, by detecting patterns in environmental data that we cannot see, since the data are so abundant and so complex. It can also help us with solutions, for example, by helping us to deal with coordination complexity and implementing measures such as cuts in harmful emissions, as Floridi et al. (2018) have argued. More generally, AI could help by means of monitoring and modeling environmental systems and by enabling solutions such as smart energy grids and smart agriculture, as a World Economic Forum blog proposes (Herweijer 2018). Governments but also companies can take the lead here. For example, Google has already used AI to cut back on data center energy use.

However, this does not necessarily "save the planet." AI can also cause problems and potentially make things worse. Consider again the negative environmental impact AI can have given the energy, infrastructures, and materials it relies on. We need to consider not only use but also production: the electricity may be produced in nonsustainable ways, and the production of AI devices uses energy and raw materials and produces waste. Or consider the "self-nudging" proposed by Floridi et al: they suggest that AI may help us to behave in environmentally good ways by helping us to stick to our self-imposed choice. But this has its own ethical risks: it is not clear that it respects human autonomy and dignity, as the authors claim, and it might go in the direction of the benign AI that takes care of humans but destroys human freedom and contributes to the problem of the Anthropocene. There is at least the *risk* of new forms of paternalism and authoritarianism. Furthermore, using AI to tackle climate change may go hand in hand with a worldview that turns the world into a mere repository of data and a view of the human that reduces human intelligence to data processing—perhaps a rather inferior kind of data processing that requires enhancement by machines. Such views are unlikely to reshape our relation to the environment in a way that mitigates challenges such as climate change and the problems indicated by the term Anthropocene.

We also face a risk of techno-solutionism in the sense that proposals for using AI to tackle environmental problems may assume that there can be a final solution to all problems, that technology alone can give the answer to our hardest questions, and that we can solve the problems entirely by use of human or artificial intelligence. But environmental problems cannot be solved entirely by means of technological-scientific intelligence; they are also linked to political and social problems that cannot be dealt with solely by means of technology. Environmental problems are always also human problems. And mathematics and its technological offspring are very helpful tools, but are limited when it comes understanding human problems and dealing with them. For example, values may conflict. AI doesn't necessarily help us answer the question about priorities, which is an important ethical and political question we should leave to humans to answer. And the humanities and social sciences teach us to be very cautious about "final" solutions.

Furthermore, humans are not the only ones with problems; nonhumans face difficulties as well, which are often neglected in discussions about the future of AI. Finally, the view that we should escape the Earth, or the worldview that everything is data that we humans can manipulate with the help of machines, could lead to a widening of the gap between the rich and poor and to large-scale forms of exploitation and violations of human dignity, as well as threatening the lives of future generations by risking to destroy the conditions for life on our planet. We need to reflect more deeply on how to build sustainable societies and environments—we need *human* thinking.

Wanted: Intelligence and Wisdom

Yet the way humans think also has various sides. AI is related to one kind of human thinking and intelligence: the more abstract, cognitive kind. This kind of thinking has proved very successful, but it has its limitations and is not the only kind of thinking we can or should do. Answering ethical and political questions concerning how to live, how to deal with our environment, and how we best relate to nonhuman living beings requires more than abstract human intelligence (e.g., arguments, theories, models) or AI pattern recognition. We need smart people and intelligent machines, but we also need intuitions and know-how that cannot be made entirely explicit, and we need to develop practical wisdom and virtue in response to concrete problems and situations and in order to decide our priorities. Such wisdom may well be informed by abstract cognitive processes and data analysis, but it is also based on embodied, relational, and situational experiences in the world, on dealing with other people, with materiality, and with our natural environment. Our success in tackling the big problems of our time will most likely depend on combinations of abstract intelligence—human and artificial—and concrete practical wisdom developed on the basis of concrete and situational human experience and practice—including our experience with technology. In whatever direction the further development of AI goes, the challenge to develop the latter kind of knowledge and learning is ours. Humans have to do it. AI is good at recognizing patterns, but wisdom cannot be delegated to machines.

Anthropocene
The alleged current geological epoch in which humanity has dramatically increased its power and effect on the Earth and its ecosystems, turning humans into a geological force.

Artificial intelligence (AI)
Intelligence displayed or simulated by technological means. Often it is assumed that "intelligence" in this definition means: considered intelligent by the standard of human intelligence, the sort of intelligent capacities and behavior that humans display. The term can refer to the science or to the technologies, for example, learning algorithms.

Bias
Discrimination against or in favor of particular individuals or groups. In the context of ethics and politics, the question arises whether a particular bias is unjust or unfair.

Data science
Interdisciplinary science that uses statistics, algorithms, and other methods to extract meaningful and useful patterns from data sets—sometimes known as "big data." Today, *machine learning* is often used in this field. Next to analysis of data, data science is also concerned with the capturing, preparation, and interpretation of data.

Deep learning
A form of *machine learning* that uses neural networks with several layers of "neurons": simple interconnected processing units that interact.

Ethics by design
An approach to technology ethics and a key component of *responsible innovation* that aims to integrate ethics in the design and development stage of the technology. Sometimes formulated as "embedding values in design." Similar terms are "value-sensitive design" and "ethically aligned design."

Explainability
The ability to explain or be explained. In the context of ethics, it refers to the ability to explain to others why you have done something or why you have made a decision; this is part of what it means to be responsible.

Explainable AI
AI that can explain to humans its actions, decisions, or recommendations, or can provide sufficient information about how it came to its result.

General AI
Human-like intelligence, which can be applied widely as opposed to narrow AI, which can only be applied to one particular problem or task. Also called "strong" AI as opposed to "weak" AI.

Machine learning
A machine or software that can automatically learn: not in the way humans learn, but based on a computational and statistical process. Feeding on data, learning algorithms can detect patterns or rules in the data and make predictions for future data.

Moral agency
The capacity for moral action, reasoning, judgment, and decision making, as opposed to merely having moral consequences.

Moral patiency
The moral standing of an entity in the sense of how that entity should be treated.

Moral responsibility
Can be used as a synonym for "being moral" and will then refer to producing morally good consequences, adhering to moral principles, being virtuous, deserving praise, and so on—the emphasis depending on the normative theory assumed. One can also ask under what conditions one can be held responsible. Conditions for attributing moral responsibility are *moral agency* and knowledge. Relational approaches stress that one is always answerable to others.

Positive ethics
Ethics concerned with how we should live (together), based on a vision of the good life and the good society. Contrasts with negative ethics, which sets limits and says what we should not do.

Posthumanism
A range of beliefs that questions humanism, especially the central position of the human being, and expands the circle of ethical concern to nonhumans.

Responsible innovation
Approach to rendering innovation more ethically and societally responsible, which typically entails embedding ethics in design and taking into account the opinions and interests of stakeholders.

Superintelligence
The idea that machines will surpass human intelligence. Sometimes connected with the idea of an "intelligence explosion" caused by intelligent machines designing even more intelligent machines.

Sustainable AI
AI that enables and contributes to a sustainable way of living for humanity and does not destroy the ecosystems on Earth on which humans (and also many nonhumans) depend.

Symbolic AI
AI that relies on symbolic representations of higher cognitive tasks such as abstract reasoning and decision making. It may use a decision tree and take the form of an expert system that requires input from domain experts.

Technological singularity
The idea that there will be a moment in human history when an explosion of machine intelligence will bring such a dramatic change to our civilization that we will no longer understand what is happening.

Transhumanism
The belief that humans should enhance themselves by means of advanced technologies and in this way transform the human condition: humanity should move to a next stage. It is also an international movement.

Trustworthy AI
AI that can be trusted by humans. Conditions for such trust can refer to (other) ethical principles such as human dignity, respect for human rights, and so on, and/or to social and technical factors that influence whether people will want to use the technology. The use of the term "trust" with regard to technologies is controversial.

NOTES

Chapter 1
1. See https://www.youtube.com/watch?v=D5VN56jQMWM.
2. See the case of Paul Zilly as told by Fry (2018, 71–72). More details in Julia Angwin, Jeff Larson, Surya Mattu and Lauren Kirchner, "Machine Bias," ProPublica, May 23, 2016, https://www.propublica.org/article/machine-bias-risk-assessments-in-criminal-sentencing.
3. For example, in 2016 a local police zone in Belgium started using predictive policing software to predict burglaries and vehicle theft (AlgorithmWatch 2019, 44).
4. BuzzFeedVideo, "You Won't Believe What Obama Says in This Video!" https://www.youtube.com/watch?v=cQ54GDm1eL0&fbclid=IwAR1oD0Alop EZa00XIIo3WNcey_qNnNqTsvHN_aZsNb0d2t9cmsDbm9oCfX8A.

Chapter 2
1. Some talk of taming or domesticating AI, although the analogy with wild animals is problematic, if only because in contrast to the "wild" AI some imagine, animals are limited by their natural faculties and can be trained and developed only up to some point (Turner 2019).
2. It is often suggested that Mary Shelley must have been influenced by her parents, who discussed politics, philosophy, and literature, but also science, and by her partner Percy Bysshe Shelley, who was an amateur scientist especially interested in electricity.

Chapter 3
1. Dreyfus was influenced by Edmund Husserl, Martin Heidegger, and Maurice Merleau-Ponty.

Chapter 4
1. A real-world case of this was the robot dog Spot who was kicked by its developers to test it, something that met with surprisingly empathetic responses: https://www.youtube.com/watch?v=aR5Z6AoMh6U.

Chapter 5
1. See https://www.humanbrainproject.eu/en/.

2. See, for example, the European Commission's AI High Level Expert Group's (2018) definition of AI.

Chapter 6
1. See http://tylervigen.com/spurious-correlations.
2. Concrete examples such as Facebook, Walmart, American Express, Hello Barbie, and BMW are drawn from Marr (2018).

Chapter 8
1. One could ask, however, if decisions made by AIs really count as decisions, and if so, if there is a difference in the kind of decisions we delegate or should delegate to AIs. In this sense, the problem regarding responsibility of or for AI raises the very question of what a decision is. The problem also connects with issues about delegation: we delegate decisions to machines. But what does this delegation entail in terms of responsibility?
2. Indeed, this case is more complicated since one could argue that the delegate is then still responsible for that particular task—at least to some extent— and it may not be clear how the responsibility is distributed in such cases.
3. Note that this was and is not always the case; as Turner (2019) reminds us, there are cases of animals being punished.

Chapter 9
1. Thanks to Bill Price for the thought experiment.

Chapter 10
1. See https://www.acrai.at/en/.
2. The resolution can be found here: http://www.europarl.europa.eu/doceo/document/TA-8-2017-0051_EN.html?redirect#title1.
3. See https://www.scu.edu/ethics-in-technology-practice/conceptual-frameworks/.
4. See https://www.partnershiponai.org/.
5. See https://www.blog.google/technology/ai/ai-principles/.
6. See https://www.microsoft.com/en-us/ai/our-approach-to-ai.
7. See https://www.accenture.com/t20160629T012639Z__w__/us-en/_acnmedia/PDF-24/Accenture-Universal-Principles-Data-Ethics.pdf.
8. See https://www.businessinsider.de/apple-ceo-tim-cook-on-privacy-the-free-market-is-not-working-regulations-2018-11?r=US&IR=T.
9. See https://leginfo.legislature.ca.gov/faces/billTextClient.xhtml?bill_id=201720180SB1001.

10. See https://www.stopkillerrobots.org/.

11. See https://futureoflife.org/ai-principles/.

12. Consider people such as Batya Friedman and Helen Nissenbaum in the United States, and later Jeroen van den Hoven and others in the Netherlands, who have been championing the ethical design of technology for some time.

13. See https://www.tuev-sued.de/company/press/press-archive/tuv-sud -and-dfki-to-develop-tuv-for-artificial-intelligence.

Chapter 11

1. See https://ec.europa.eu/digital-single-market/en/european-ai-alliance.

Chapter 12

1. See https://hai.stanford.edu/ and https://hcai.mit.edu.

2. See https://sustainabledevelopment.un.org/post2015/transforming ourworld.

3. See https://www.theguardian.com/science/2018/feb/07/space-oddity -elon-musk-spacex-car-mars-falcon-heavy.

4. See https://cosmosmagazine.com/space/why-we-need-to-send-artists -into-space.

REFERENCES

Accessnow. 2018. "Mapping Regulatory Proposals for Artificial Intelligence in Europe." https://www.accessnow.org/cms/assets/uploads/2018/11/mapping_regulatory_proposals_for_AI_in_EU.pdf.

ACRAI (Austria Council on Robotics and Artificial Intelligence). 2018. "Die Zukunft Österreichs mit Robotik und Künstlicher Intelligenz positive gestalten: White paper des Österreichischen Rats für Robotik und Künstliche Intelligenz."

"Algorithm and Blues." 2016. *Nature* 537:449.

AlgorithmWatch. 2019. "Automating Society: Taking Stock of Automated Decision Making in the EU." A report by AlgorithmWatch in cooperation with Bertelsmann Stiftung. January 2019. Berlin: AW AlgorithmWatch GmbH. http://www.algorithmwatch.org/automating-society.

Alpaydin, Ethem. 2016. *Machine Learning*. Cambridge, MA: MIT Press.

Anderson, Michael and Susan Anderson. 2011. "General Introduction." In *Machine Ethics*, edited by Michael Anderson and Susan Anderson, 1–4. Cambridge: Cambridge University Press.

Arendt, Hannah. 1958. *The Human Condition*. Chicago: Chicago University Press.

Arkoudas, Konstantine, and Selmer Bringsjord. 2014. "Philosophical Foundations." In *The Cambridge Handbook of Artificial Intelligence*, edited by Keith Frankish and William M. Ramsey. Cambridge: Cambridge University Press.

Armstrong, Stuart. 2014. *Smarter Than Us: The Rise of Machine Intelligence*. Berkeley: Machine Intelligence Research Institute.

Awad, Edmond, Sohan Dsouza, Richard Kim, Jonathan Schulz, Joseph Henrich, Azim Shariff, Jean-François Bonnefon, and Iyad Rahwan. 2018. "The Moral Machine Experiment." *Nature* 563:59–64.

Bacon, Francis. 1964. "The Refutation of Philosophies." In *The Philosophy of Francis Bacon*, edited by Benjamin Farrington, 103–132. Chicago: University of Chicago Press.

Boddington, Paula. 2016. "The Distinctiveness of AI Ethics, and Implications for Ethical Codes." Paper presented at the workshop Ethics for Artificial Intelligence, July 9, 2016, IJCAI-16, New York. https://www.cs.ox.ac.uk/efai/2016/11/02/the-distinctiveness-of-ai-ethics-and-implications-for-ethical-codes/.

Boddington, Paula. 2017. *Towards a Code of Ethics for Artificial Intelligence*. Cham: Springer.

Boden, Margaret A. 2016. *AI: Its Nature and Future*. Oxford: Oxford University Press.

Borowiec, Steven. 2016. "AlphaGo Seals 4–1 Victory Over Go Grandmaster Lee Sedol." *Guardian*, March 15. https://www.theguardian.com/technology/2016/mar/15/googles-alphago-seals-4-1-victory-over-grandmaster-lee-sedol.

Bostrom, Nick. 2014. *Superintelligence*. Oxford: Oxford University Press.

Brynjolfsson, Erik, and Andrew McAfee. 2014. *The Second Machine Age*. New York: W. W. Norton.

Bryson, Joanna. 2010. "Robots Should Be Slaves." In *Close Engagements with Artificial Companions: Key Social, Psychological, Ethical and Design Issues*, edited by Yorick Wilks, 63–74. Amsterdam: John Benjamins.

Bryson, Joanna. 2018. "AI & Global Governance: No One Should Trust AI." United Nations University Centre for Policy Research. *AI & Global Governance*, November 13, 2018. https://cpr.unu.edu/ai-global-governance-no-one-should-trust-ai.html.

Bryson, Joanna, Mihailis E. Diamantis, and Thomas D. Grant. 2017. "Of, For, and By the People: The Legal Lacuna of Synthetic Persons." *Artificial Intelligence & Law* 25, no. 3: 273–291.

Caliskan, Aylin, Joanna J. Bryson, and Arvind Narayanan. 2017. "Semantics Derived Automatically from Language Corpora Contain Human-like Biases." *Science* 356:183–186.

Castelvecchi, Davide. 2016. "Can We Open the Black Box of AI?" *Nature* 538, no. 7623: 21–23.

CDT (Centre for Democracy & Technology) 2018. "Digital Decisions." https://cdt.org/issue/privacy-data/digital-decisions/.

Coeckelbergh, Mark. 2010. "Moral Appearances: Emotions, Robots, and Human Morality." *Ethics and Information Technology* 12, no. 3: 235–241.

Coeckelbergh, Mark. 2011. "You, Robot: On the Linguistic Construction of Artificial Others." *AI & Society* 26, no. 1: 61–69.

Coeckelbergh, Mark. 2012. *Growing Moral Relations: Critique of Moral Status Ascription*. New York: Palgrave Macmillan.

Coeckelbergh, Mark. 2013. *Human Being @ Risk: Enhancement, Technology, and the Evaluation of Vulnerability Transformations*. Cham: Springer.

Coeckelbergh, Mark. 2017. *New Romantic Cyborgs*. Cambridge, MA: MIT Press.

Crawford, Kate, and Ryan Calo. 2016. "There Is a Blind Spot in AI Research." *Nature* 538:311–313.

Crutzen, Paul J. 2006. "The 'Anthropocene.'" In *Earth System Science in the Anthropocene* edited by Eckart Ehlers and Thomas Krafft, 13–18. Cham: Springer.

Darling, Kate, Palash Nandy, and Cynthia Breazeal. 2015. "Empathic Concern and the Effect of Stories in Human-Robot Interaction." In *2015 24th IEEE International Symposium on Robot and Human Interactive Communication (RO-MAN)*, 770–775. New York: IEEE.

Dennett, Daniel C. 1997. "Consciousness in Human and Robot Minds. In *Cognition, Computation, and Consciousness*, edited by Masao Ito, Yasushi Miyashita, and Edmund T. Rolls, 17–29. New York: Oxford University Press.

Digital Europe. 2018. "Recommendations on AI Policy: Towards a Sustainable and Innovation-friendly Approach." Digitaleurope.org, November 7, 2018.

Dignum, Virginia, Matteo Baldoni, Cristina Baroglio, Maruiyio Caon, Raja Chatila, Louise Dennis, Gonzalo Génova, et al. 2018. "Ethics by Design: Necessity or Curse?" Association for the Advancement of Artificial Intelligence. http://www.aies-conference.com/2018/contents/papers/main/AIES_2018 _paper_68.pdf.

Dowd, Maureen. 2017. "Elon Musk's Billion-Dollar Crusade to Stop the A.I. Apocalypse." *Vanity Fair*, March 26, 2017. https://www.vanityfair.com/news/2017/03/elon-musk-billion-dollar-crusade-to-stop-ai-space-x.

Dreyfus, Hubert L. 1972. *What Computers Can't Do*. New York: HarperCollins.

Druga, Stefania and Randi Williams. 2017. "Kids, AI Devices, and Intelligent Toys." MIT Media Lab, June 6, 2017. https://www.media.mit.edu/posts/kids-ai-devices/f.

European Commission. 2018. "Ethics and Data Protection." http://ec.europa.eu/research/participants/data/ref/h2020/grants_manual/hi/ethics/h2020_hi_ethics-data-protection_en.pdf.

European Commission Directorate-General of Employment, Social Affairs and Inclusion. 2018. "Employment and Social Developments in Europe 2018." Luxembourg: Publications Office of the European Union. http://ec.europa.eu/social/main.jsp?catId=738&langId=en&pubId=8110.

European Commission AI HLEG (High-Level Expert Group on Artificial Intelligence). 2018. "Draft Ethics Guidelines for Trustworthy AI: Working Document for Stakeholders." Working document, December 18, 2018. Brussels: European Commission. https://ec.europa.eu/digital-single-market/en/news/draft-ethics-guidelines-trustworthy-ai.

European Commission AI HLEG (High-Level Expert Group on Artificial Intelligence). 2019. "Ethics Guidelines for Trustworthy AI." April 8, 2019. Brussels: European Commission. https://ec.europa.eu/futurium/en/ai-alliance-consultation/guidelines#Top.

EGE (European Group on Ethics in Science and New Technologies). 2018. "Statement on Artificial Intelligence, Robotics and 'Autonomous' Systems." Brussels: European Commission.

European Parliament and the Council of the European Union. 2016. "General Data Protection Regulation (GDPR)." https://eur-lex.europa.eu/legal-content/EN/TXT/?uri=celex%3A32016R0679.

Executive Office of the President, National Science and Technology Council Committee on Technology. 2016. "Preparing for the Future of Artificial Intelligence." Washington, DC: Office of Science and Technology Policy (OSTP).

Floridi, Luciano, Josh Cowls, Monica Beltrametti, Raja Chatila, Patrice Chazerand, Virginia Dignum, Christoph Luetge, Robert Madelin, Ugo Pagallo, Francesca Rossi, Burkhard Schafer, Peggy Valcke, and Effy Vayena. 2018. "AI4People—An Ethical Framework for a Good AI Society: Opportunities, Risks, Principles, and Recommendations." *Minds and Machines* 28, no. 4: 689–707.

Floridi, Luciano, and J. W. Sanders. 2004. "On the Morality of Artificial Agents." *Minds and Machines* 14, no. 3: 349–379.

Ford, Martin. 2015. *Rise of the Robots: Technology and the Threat of a Jobless Future*. New York: Basic Books.

Frankish, Keith, and William M. Ramsey. 2014. "Introduction." In *The Cambridge Handbook of Artificial Intelligence*, edited by Keith Frankish and William M. Ramsey, 1–14. Cambridge: Cambridge University Press.

Frey, Carl Benedikt, and Michael A. Osborne. 2013. "The Future of Employment: How Susceptible Are Jobs to Computerisation?" Working paper, Oxford Martin Programme on Technology and Employment, University of Oxford.

Fry, Hannah. 2018. *Hello World: Being Human in the Age of Algorithms*. New York: W. W. Norton.

Fuchs, Christian. 2014. *Digital Labour and Karl Marx*. New York: Routledge.

Goebel, Randy, Ajay Chander, Katharina Holzinger, Freddy Lecue, Zeynep Akata, Simone Stumpf, Peter Kieseberg, and Andreas Holzinger. 2018. "Explainable AI: The New 42?" Paper presented at the CD-MAKE 2018, Hamburg, Germany, August 2018.

Gunkel, David. 2012. *The Machine Question*. Cambridge, MA: MIT Press.

Gunkel, David. 2018. "The Other Question: Can and Should Robots Have Rights?" *Ethics and Information Technology* 20:87–99.

Harari, Yuval Noah. 2015. *Homo Deus: A Brief History of Tomorrow*. London: Hervill Secker.

Haraway, Donna. 1991. "A Cyborg Manifesto: Science, Technology, and Socialist-Feminism in the Late Twentieth Century." In *Simians, Cyborgs and Women: The Reinvention of Nature*, 149–181. New York: Routledge.

Haraway, Donna. 2015. "Anthropocene, Capitalocene, Plantationocene, Chthulucene: Making Kin." *Environmental Humanities* 6:159–165.

Herweijer, Celine. 2018. "8 Ways AI Can Help Save the Planet." *World Economic Forum*, January 24, 2018. https://www.weforum.org/agenda/2018/01/8-ways-ai-can-help-save-the-planet/.

House of Commons. 2018. "Algorithms in Decision-Making." Fourth Report of Session 2017-19, HC351. May 23, 2018.

ICDPPC (International Conference of Data Protection and Privacy Commissioners). 2018. "Declaration on Ethics and Data Protection in Artificial Intelligence." https://icdppc.org/wp-content/uploads/2018/10/20180922_ICDPPC-40th_AI-Declaration_ADOPTED.pdf.

IEEE Global Initiative on Ethics of Autonomous and Intelligent Systems. 2017. "Ethically Aligned Design: A Vision for Prioritizing Human Well-Being with Autonomous and Intelligent Systems," Version 2. IEEE. http://standards.Ieee.org/develop/indconn/ec/autonomous_systems.html.

Ihde, Don. 1990. *Technology and the Lifeworld: From Garden to Earth*. Bloomington: Indiana University Press.

Jansen, Philip, Stearns Broadhead, Rowena Rodrigues, David Wright, Philp Brey, Alice Fox, and Ning Wang. 2018. "State-of-the-Art Review." Draft of the D4.1 deliverable submitted to the European Commission on April 13, 2018. A report for The SIENNA Project, an EU H2020 research and innovation program under grant agreement no. 741716.

Johnson, Deborah G. 2006. "Computer Systems: Moral Entities but not Moral Agents." *Ethics and Information Technology* 8, no. 4: 195–204.

Kant, Immanuel. 1997. *Lectures on Ethics*. Edited by Peter Heath and J. B. Schneewind. Translated by Peter Heath. Cambridge: Cambridge University Press.

Kelleher, John D., and Brendan Tierney. 2018. *Data Science*. Cambridge, MA: MIT Press.

Kharpal, Arjun. 2017. "Stephen Hawking Says A.I. Could Be 'Worst Event in the History of Our Civilization.'" CNBC. November 6, 2017. https://www.cnbc.com/2017/11/06/stephen-hawking-ai-could-be-worst-event-in-civilization.html.

Kubrick, Stanley, dir. 1968. *2001: A Space Odyssey*. Beverly Hills, CA: Metro-Goldwyn-Mayer.

Kurzweil, Ray. 2005. *The Singularity Is Near*. New York: Viking.

Leta Jones, Meg. 2018. "Silencing Bad Bots: Global, Legal and Political Questions for Mean Machine Communication." *Communication Law and Policy* 23, no. 2: 159–195.

Lin, Patrick, Keith Abney, and George Bekey. 2011. "Robot Ethics: Mapping the Issues for a Mechanized World." *Artificial Intelligence* 175:942–949.

MacIntyre, Lee C. 2018. *Post-Truth*. Cambridge, MA: MIT Press.

Marcuse, Herbert. 1991. *One-Dimensional Man*. Boston: Beacon Press.

Marr, Bernard. 2018. "27 Incredible Examples of AI and Machine Learning in Practice." *Forbes*, April 30. https://www.forbes.com/sites/bernardmarr/2018/04/30/27-incredible-examples-of-ai-and-machine-learning-in-practice/#6b37edf27502.

McAfee, Andrew, and Erik Brynjolfsson. 2017. *Machine, Platform, Crowd: Harnessing Our Digital Future*. New York: W. W. Norton.

Miller, Tim. 2018. "Explanation in Artificial Intelligence: Insights from the Social Sciences." *arXiv*, August 15. https://arxiv.org/pdf/1706.07269.pdf.

Mouffe, Chantal. 2013. *Agonistics: Thinking the World Politically*. London: Verso.

Nemitz, Paul Friedrich, 2018. "Constitutional Democracy and Technology in the Age of Artificial Intelligence." *Philosophical Transactions of the Royal Society A* 376, no. 2133. https://doi.org/10.1098/rsta.2018.0089.

Noble, David F. 1997. *The Religion of Technology*. New York: Penguin Books.

Reijers, Wessel, David Wright, Philip Brey, Karsten Weber, Rowena Rodrigues, Declan O' Sullivan, and Bert Gordijn. 2018. "Methods for Practising Ethics in Research and Innovation: A Literature Review, Critical Analysis and Recommendation." *Science and Engineering Ethics* 24, no. 5: 1437–1481.

Royal Society, the. 2018. "Portrayals and Perceptions of AI and Why They Matter." December 11, 2018. https://royalsociety.org/topics-policy/projects/ai-narratives/.

Rushkoff, Douglas. 2018. "Survival of the Richest." *Medium*, July 5. https://medium.com/s/futurehuman/survival-of-the-richest-9ef6cddd0cc1.

Samek, Wojciech, Thomas Wiegand, and Klaus-Robert Müller. 2017. "Explainable Artificial Intelligence: Understanding, Visualizing and Interpreting Deep Learning Models." https://arxiv.org/pdf/1708.08296.pdf.

Schwab, Katharine. 2018. "The Exploitation, Injustice, and Waste Powering Our AI." *Fast Company*. September 18, 2018. https://www.fastcompany.com/90237802/the-exploitation-injustice-and-waste-powering-our-ai.

Seseri, Rudina. 2018. "The Problem with 'Explainable AI.'" *Tech Crunch*. June 14, 2018. https://techcrunch.com/2018/06/14/the-problem-with-explainable-ai/?guccounter=1

Searle, John. R. 1980. "Minds, Brains, and Programs." *Behavioral and Brain Sciences* 3, no. 3: 417–457.

Shanahan, Murray. 2015. *The Technological Singularity*. Cambridge, MA: The MIT Press.

Siau, Keng, and Weiyu Wang. 2018. "Building Trust in Artificial Intelligence, Machine Learning, and Robotics." *Cutter Business Technology Journal* 32, no. 2: 46–53.

State Council of China. 2017. "New Generation Artificial Intelligence Development Plan." Translated by Flora Sapio, Weiming Chen, and Adrian Lo. https://flia.org/notice-state-council-issuing-new-generation-artificial-intelligence-development-plan/.

Stoica, Ion. 2017. "A Berkeley View of Systems Challenges for AI." Technical Report No. UCB/EECS-2017-159. http://www2.eecs.berkeley.edu/Pubs/TechRpts/2017/EECS-2017.

Sullins, John. 2006. "When Is a Robot a Moral Agent?" *International Review of Information Ethics* 6: 23–30.

Surur. 2017. "Microsoft Aims to Lie to Their AI to Reduce Sexist Bias." August 25, 2017. https://mspoweruser.com/microsoft-aims-lie-ai-reduce-sexist-bias/.

Suzuki, Yutaka, Lisa Galli, Ayaka Ikeda, Shoji Itakura, and Michiteru Kitazaki. 2015. "Measuring Empathy for Human and Robot Hand Pain Using Electroencephalography." *Scientific Reports* 5, article number 15924. https://www.nature.com/articles/srep15924

Tegmark, Max. 2017. *Life 3.0: Being Human in the Age of Artificial Intelligence*. Allen Lane/Penguin Books.

Turkle, Sherry. 2011. *Alone Together: Why We Expect More from Technology and Less from Each Other*. New York: Basic Books.

Turner, Jacob. 2019. *Robot Rules: Regulating Artificial Intelligence*. Cham: Palgrave Macmillan.

Université de Montréal. 2017. "Montréal Declaration Responsible AI." https://www.montrealdeclaration-responsibleai.com/the-declaration.

Vallor, Shannon. 2016. *Technology and the Virtues*. New York: Oxford University Press.

Vigen, Tyler. 2015. *Spurious Correlations*. New York: Hachette Books.

Villani, Cédric. 2018. *For a Meaningful Artificial Intelligence: Towards a French and European Strategy*. Composition of a parliamentary mission from September 8, 2017, to March 8, 2018, and assigned by the Prime Minister of France, Èdouard Philippe.

Von Schomberg, René, ed. 2011. "Towards Responsible Research and Innovation in the Information and Communication Technologies and Security Technologies Fields." A report from the European Commission Services. Luxembourg: Publications Office of the European Union.

Vu, Mai-Anh T., Tülay Adalı, Demba Ba, György Buzsáki, David Carlson, Katherine Heller, et al. 2018. "A Shared Vision for Machine Learning in Neuroscience." *Journal of Neuroscience* 38, no. 7: 1601–607.

Wachter, Sandra, Brent Mittelstadt, and Luciano Floridi. 2017. "Why a Right to Explanation of Automated Decision-Making Does Not Exist in the General Data Protection Regulation." *International Data Privacy Law, 2017*. http://dx.doi.org/10.2139/ssrn.2903469.

Wallach, Wendell and Colin Allen. 2009. *Moral Machines: Teaching Robots Right from Wrong*. Oxford: Oxford University Press.

Weld, Daniel S. and Gagan Bansal. 2018. "The Challenge of Crafting Intelligible Intelligence." https://arxiv.org/pdf/1803.04263.pdf.

Winfield, Alan F.T. and Marina Jirotka. 2017. "The Case for an Ethical Black Box." In *Towards Autonomous Robotic Systems*, edited by Yang Gao, Saber Fallah, Yaochu Jin, and Constantina Lekakou (proceedings of TAROS 2017, Guildford, UK, July 2017), 262–273. Cham: Springer.

Winikoff, Michael. 2018. "Towards Trusting Autonomous Systems." In *Engineering Multi-Agent Systems*, edited by Amal El Fallah Seghrouchni, Alessandro Ricci, and Son Trao, 3–20. Cham: Springer.

Yampolskiy, Roman V. 2013. "Artificial Intelligence Safety Engineering: Why Machine Ethics Is a Wrong Approach." In *Philosophy and Theory of Artificial Intelligence* edited by Vincent C. Müller, 289–296. Cham: Springer.

Yeung, Karen. 2018. "A Study of the Implications of Advanced Digital Technologies (Including AI Systems) for the Concept of Responsibility within a Human Rights Framework." A study commissioned for the Council of Europe Committee of experts on human rights dimensions of automated data processing and different forms of artificial intelligence. MSI-AUT (2018)05.

Zimmerman, Jess. 2015. "What If the Mega-Rich Just Want Rocket Ships to Escape the Earth They Destroy?" *Guardian*, September 16, 2015. https://www.theguardian.com/commentisfree/2015/sep/16/mega-rich-rocket-ships-escape-earth.

Zou, James, and Londa Schiebinger. 2018. "Design AI So That It's Fair." *Nature* 559:324–326.

FURTHER READING

Alpaydin, Ethem. 2016. *Machine Learning*. Cambridge, MA: MIT Press.

Arendt, Hannah. 1958. *The Human Condition*. Chicago: Chicago University Press.

Aristotle. 2002. *Nichomachean Ethics*. Translated by Christopher Rowe, with commentary by Sarah Broadie. Oxford: Oxford University Press.

Boddington, Paula. 2017. *Towards a Code of Ethics for Artificial Intelligence*. Cham: Springer.

Boden, Margaret A. 2016. *AI: Its Nature and Future*. Oxford: Oxford University Press.

Bostrom, Nick. 2014. *Superintelligence*. Oxford: Oxford University Press.

Brynjolfsson, Erik, and Andrew McAfee. 2014. *The Second Machine Age*. New York: W.W. Norton.

Coeckelbergh, Mark. 2012. *Growing Moral Relations: Critique of Moral Status Ascription*. New York: Palgrave Macmillan.

Crutzen, Paul J. 2006. "The 'Anthropocene.'" In *Earth System Science in the Anthropocene*, edited by Eckart Ehlers and Thomas Krafft, 13–18. Cham: Springer.

Dignum, Virginia, Matteo Baldoni, Cristina Baroglio, Maruiyio Caon, Raja Chatila, Louise Dennis, Gonzalo Génova, et al. 2018. "Ethics by Design: Necessity or Curse?" Association for the Advancement of Artificial Intelligence. http://www.aies-conference.com/2018/contents/papers/main/AIES_2018 _paper_68.pdf.

Dreyfus, Hubert L. 1972. *What Computers Can't Do*. New York: Harper & Row.

Floridi, Luciano, Josh Cowls, Monica Beltrametti, Raja Chatila, Patrice Chazerand, Virginia Dignum, Christoph Luetge, Robert Madelin, Ugo Pagallo, Francesca Rossi, Burkhard Schafer, Peggy Valcke, and Effy Vayena. 2018.

"AI4People—An Ethical Framework for a Good AI Society: Opportunities, Risks, Principles, and Recommendations." *Minds and Machines* 28, no. 4: 689–707.

Frankish, Keith, and William M. Ramsey, eds. 2014. *The Cambridge Handbook of Artificial Intelligence*. Cambridge: Cambridge University Press.

European Commission AI HLEG (High-Level Expert Group on Artificial Intelligence). 2019. "Ethics Guidelines for Trustworthy AI." April 8, 2019. Brussels: European Commission. https://ec.europa.eu/futurium/en/ai-alliance -consultation/guidelines#Top.

Fry, Hannah. 2018. *Hello World: Being Human in the Age of Algorithms*. New York and London: W. W. Norton.

Fuchs, Christian. 2014. *Digital Labour and Karl Marx*. New York: Routledge.

Gunkel, David. 2012. *The Machine Question*. Cambridge, MA: MIT Press.

Harari, Yuval Noah. 2015. *Homo Deus: A Brief History of Tomorrow*. London: Hervill Secker.

Haraway, Donna. 1991. "A Cyborg Manifesto: Science, Technology, and Socialist-Feminism in the Late Twentieth Century." In *Simians, Cyborgs and Women: The Reinvention of Nature*, 149–181. New York: Routledge.

IEEE Global Initiative on Ethics of Autonomous and Intelligent Systems. 2017. "Ethically Aligned Design: A Vision for Prioritizing Human Well-being with Autonomous and Intelligent Systems," Version 2. IEEE, 2017. http:// standards.Ieee.org/develop/indconn/ec/autonomous_systems.html.

Kelleher, John D. and Brendan Tierney. 2018. *Data Science*. Cambridge, MA: MIT Press.

Nemitz, Paul Friedrich, 2018. "Constitutional Democracy and Technology in the Age of Artificial Intelligence." *Philosophical Transactions of the Royal Society A* 376, no. 2133. https://doi.org/10.1098/rsta.2018.0089

Noble, David F. 1997. *The Religion of Technology*. New York: Penguin Books.

Reijers, Wessel, David Wright, Philip Brey, Karsten Weber, Rowena Rodrigues, Declan O'Sullivan, and Bert Gordijn. 2018. "Methods for Practising Ethics in Research and Innovation: A Literature Review, Critical Analysis and Recommendation." *Science and Engineering Ethics* 24, no. 5: 1437–1481.

Shelley, Mary. 2017. *Frankenstein*. Annotated edition. Edited by David H. Guston, Ed Finn, and Jason Scott Robert. Cambridge, MA: MIT Press.

Turkle, Sherry. 2011. *Alone Together: Why We Expect More from Technology and Less from Each Other*. New York: Basic Books.

Wallach, Wendell, and Colin Allen. 2009. *Moral Machines: Teaching Robots Right from Wrong*. Oxford: Oxford University Press.

INDEX

The MIT Press Essential Knowledge Series

MARK COECKELBERGH is a full Professor of Philosophy of Media and Technology at the Philosophy of Department of the University of Vienna and the former President of the Society for Philosophy and Technology (SPT). His expertise focuses on ethics of new and emerging technologies, in particular robotics and artificial intelligence. He is currently a member of various entities that support policy building in the area of robotics and artificial intelligence, such as the European Commission's High Level Expert Group on Artificial Intelligence and the Austrian Council on Robotics and Artificial Intelligence. He is the author of twelve philosophy books and numerous articles, and is involved in several European research projects.